The Cacti of Arizona

Flowering barrel cactus in the Arizona Desert

Third edition

The Cacti
of Arizona

LYMAN BENSON

Drawings by
LUCRETIA BREAZEALE HAMILTON

THE UNIVERSITY OF ARIZONA PRESS Tucson, Arizona

About the Author

Lyman Benson has devoted his professional life to teaching and to research on the plant life of the southwestern deserts and of North America north of Mexico. As Emeritus Professor and Chairman of the Department of Botany and Director of the Herbarium of Pomona College, he has to his credit more than seventy published works on the flora of North America and especially of the Southwest. Among his books are *Trees and Shrubs of the Southwestern Deserts*, with Robert A. Darrow (University of Arizona Press), *Plant Taxonomy: Methods and Principles*, and *The Native Cacti of California*.

The University of Arizona Press
www.uapress.arizona.edu

© 1969 by The Arizona Board of Regents
All rights reserved. Published 1969
Century Collection edition 2016

Printed in the United States of America
21 20 19 18 17 16   10 9 8 7 6 5

ISBN-13: 978-0-8165-0191-5 (cloth)
ISBN-13: 978-0-8165-0509-8 (paper)
ISBN-13: 978-0-8165-3476-0 (Century Collection paper)

L. C. No. 70-77802

∞ This paper meets the requirements of ANSI/NISO Z39.48-1992 (Permanence of Paper).

Preface

The text of the third edition of *The Cacti of Arizona* is almost wholly new, being based upon new research since the second edition in 1950, which in turn was based upon more extensive field, herbarium, and bibliographic investigation than had been possible for the first edition in 1940. The pioneering investigations of the 1930s and 1940s have been eclipsed by more intensive study on all fronts and by extending research to all cactus species native in the United States and Canada and to the same and related species in Mexico and the Caribbean region. This makes possible better evaluation of the Arizona taxa in relation to those occurring elsewhere, and better understanding of the variable, geographically shifting, character combinations in populations that occur across several states or the greater part of the continent of North America.

This new edition represents part of an attempt to bring to the Cactaceae of a large geographical region a monographic study similar to those available for some other plant groups. However, despite intensive field study reaching into every state and some provinces and adjacent countries, this objective is difficult to attain. Most other plant groups are represented by extensive herbarium material, and the ramifications of species can be studied to some extent at one sitting to determine general relationships and to pinpoint problems to be studied in the field and by experimental methods. The spininess and succulence of the cacti make most collectors reluctant to secure specimens and remiss in pressing enough material to be of value. Consequently the representation of the cactus family in herbaria is perhaps only ten percent of that for most other families, and often one is fortunate to find even a single specimen of a species in any herbarium. Thus the classification of cacti is far from the precise procedure possible with, for example, *Ranunculus* (the genus of buttercups), another specialty of the writer, studied for North America in about 60 or 70 herbaria and in the field in all fifty states and most Canadian provinces. However, for *Ranunculus* the 99 North and Central American and Caribbean species are represented by at least 12 or 15 times as many herbarium specimens as the 147 species of Cactaceae in North America north of Mexico. Thus, in each of several large herbaria there are enough specimens to represent at least some elements of almost every species, and these may be compared on the spot. In the Cactaceae this is not true for any herbarium. In study of about 60 herbaria, much has been pieced together, but the herbaria are scattered over more than 6,000 miles, and nearly all information must be carried from one to another by memory and notes.

There has been progress on unraveling some of the complex and confused generic and specific problems in the Cactaceae, but many of these are far from solved. Some may remain that way for a long time to come. There are no absolute answers to problems of classification of taxa at any level — generic, specific, or varietal. Consequently, the only possible report at any time is one, hopefully, of progress. The truth is a goal, and we approach it, but it is elusive.

LYMAN BENSON

Acknowledgments

Research grants-in-aid contributing to preparation of this book have included several from the Claremont Graduate School, one from the Society of Sigma Xi for 1950 while the second edition was in press, and three from the National Science Foundation covering the periods for 1956 to 1959, 1959 to 1964, and 1965 through 1967. This aid in all phases of the work is acknowledged with gratitude.

The use of photographs furnished by several persons is much appreciated. Photographs by the late Mr. Robert H. Peebles — now property of the Herbarium of the University of Arizona under an arrangement made by the writer in 1944 for bringing the Kearney and Peebles Herbarium from Sacaton to the University — have been made available for the *Cacti of Arizona* by Dr. Charles T. Mason, Jr., of the Department of Biological Sciences. Mr. Scott E. Haselton, emeritus Editor of the *Cactus and Succulent Journal,* has granted permission for use of photographs published previously in that journal. Photographs by Drs. J. G. Brown and R. B. Streets have been made available from the files of the Department of Plant Pathology, through the courtesy of Drs. Streets and Boyle. The cooperation of the two departments is acknowledged with pleasure and gratitude. Photographs by the late David Griffiths, taken during his studies of *Opuntia* mostly from about 1903 to 1916, were made available by the United States National Herbarium, Smithsonian Institution, Washington, D.C. Photographs not attributed to other persons are those of the author.

Special appreciation is expressed to Mrs. Lucretia Breazeale Hamilton, artist for all three editions of the *Cacti of Arizona* and a joint author of an earlier University of Arizona publication, *Arizona Cacti* (1933).

The writer is grateful to the directors, curators, and staff members of about sixty college, university, and institutional herbaria where specimens have been studied.

Contents

Tables

Characters of the Varieties of:

Illustrations

List of Genera, Species, and Varieties

1. OPUNTIA, page 29

The Cacti of Arizona

Introduction

Most persons are familiar with at least one kind of cactus. However, certain other plants commonly are confused with cacti. In Arizona these others include the ocotillo; the yuccas, one of which is the Joshua tree; the sotol, or desert spoon; and the century plants, or *Agave*. These belong to several plant families, and the presence of spiny structures is almost the only characteristic they all have in common with cacti.

Distinction of all the members of the cactus family from other succulent plants requires knowledge of the outstanding characteristics of the group. Cacti may be recognized by the large, fleshy, usually leafless stems and by spines always (at least in juvenile stages) developed in clusters within spirally arranged areoles (restricted areas) on the stems. This is not true of any other plants. The cactus family is distinguished also by the following characteristics of the flowers: (1) the sepals and petals are numerous, and they intergrade with one another; (2) the pistil consists of an inferior ovary with one seed chamber with the ovules (which become seeds) on several marginal placentae and of a single style with several stigmas. These structures are explained and illustrated in the following pages.

THE STRUCTURE OF CACTI

The plant body of a cactus is fundamentally similar to that of any other flowering plant, but it is unusual in the slight development or complete elimination of leaves and the remarkable relative size of the stem. Small, fleshy, ephemeral leaves appear on the new joints of the stems of chollas and prickly pears, and well-developed persistent leaves are to be found in the primitive tropical cacti of the genus *Pereskia,* but most cacti do not have discernible leaves, or leaves are developed only on juvenile stems or at the growing point of the stem and they soon fall away. In the absence of leaves, all food manufacture is carried on by the green cells of the stems. A large part of the stem is occupied by storage cells especially adapted to water retention, and the surface is covered by a waxy epidermal layer that prevents or retards evaporation. The water-retaining power of the surface layers may be demonstrated by cutting a detached joint (segment) of a spineless prickly pear into two parts and peeling one half and leaving the other intact. After a few hours the peeled half is shriveled and the unpeeled part is unchanged.

Beneath the surface of the soil a few cacti, such as the desert night-blooming cereus, produce tuberous structures. However, the majority have shallow systems of elongated slender but fleshy roots, well adapted to the absorbing of large quantities of water during the brief periods in which it is available in the desert or arid regions.

Cactus roots remain receptive to water during prolonged dry periods. A detached joint of the stem of a prickly pear or a cholla, or the cut-off or uprooted stem of nearly any cactus, forms new roots that withstand long exposure to dry air. Thus, the wide-spreading shallow root systems are ready to absorb water whenever it comes, from even light rain or mist.

1

Fig. 0.2. Stems, spines, and leaves of a prickly pear. The stem, composed of flattened joints, has spirally arranged spine-bearing areoles. When the stem is young a leaf is present at the base of each areole. In two or three months the leaves fall off, leaving only the clusters of spines in the areoles. (Photographs by Robert H. Peebles.)

Fig. 0.1. Stems, spines, and leaves of a cholla. The stem, composed of cylindroidal joints, has spirally arranged tubercles (projections) that produce the spine-bearing areoles. When the stem is young, each tubercle bears a slender but thickened fleshy leaf at the base of each areole. In two or three months the leaf falls off, leaving only the cluster of spines in the areole. (Photographs by David Griffiths.)

The spines of some species are directed downward, and they concentrate water on their points, which, like the elongated leaf-tips of many tropical plants, serve as "drip-tips." Thus, a minor storm may be converted into waterdrops falling near the base of the cactus. In the young saguaro or giant cactus, until the stem is about three to five feet high, the principal spines are directed downward. This promotes two results: (1) concentration of water drops near the base of the young plant and (2) protection from rodents which eat cactus plants mostly for water. The curvature of the stems, especially of the joints of prickly pears, also concentrates rain into drops or even small "streams."

The detailed structure of a mature cactus is as follows:

1. The stem may be a simple, unbranched, columnar axis, as in the barrel cacti (Figs. 5.1, 5.2, 5.3), but in most cacti it is branched either at or near the base as in the organ pipe cactus (Fig. 2.6) or above the ground as in the saguaro or giant cactus (Figs. 2.1, 2.2). The stems of chollas and prickly pears consist of branching series of joints. The joints of the chollas are cylindroidal (Fig. 0.1), while those of the prickly pears are flattened (Fig. 0.2). Nearly all prickly pears have a smooth stem surface, but the stems of most other cacti have either prominent ribs like those of the saguaro (Fig. 2.5, *above*) or tubercles like those of chollas (Fig. 0.1) or

of *Mammillaria* (Fig. 4.6). Complex structures, called areoles, each formed from an axillary bud (Figs. 0.1, 0.2), are arranged spirally on the stem, one just above the potential position of each leaf. Tuberculate stems have an areole at the apex of each tubercle. Ribbed stems, like those of the saguaro and organ pipe cactus, produce areoles along the summits of the ribs or ridges formed by coalescense of the tubercles (Figs. 5.5, 5.7). The areoles give rise to clusters of spines which are, fundamentally, specialized portions of leaves. Some may be distinguished as central spines and others as marginal or radial spines (Fig. 5.9). The distinction of central and radial spines is arbitrary, and in some cases the classification is a matter of individual interpretation. Chollas and prickly pears have numerous small or minute barbed bristles as well as spines. These are called "glochids" ("ch" pronounced like "k"), and in many prickly pears they are more troublesome to humans than are the spines.

2. The flower of a cactus has an *inferior ovary;* the seed-producing vessel, or *ovary,* which becomes the fruit, is located below the conspicuous leafy or petallike flower parts and not above them, or at least not attached above them, as it is in the majority of flowering plants. A common collar of tissue, the *floral tube* (or *floral cup,* depending upon its shape), may continue for a considerable distance above the ovary, as in the night-blooming cereus, or the tube may be short or wholly

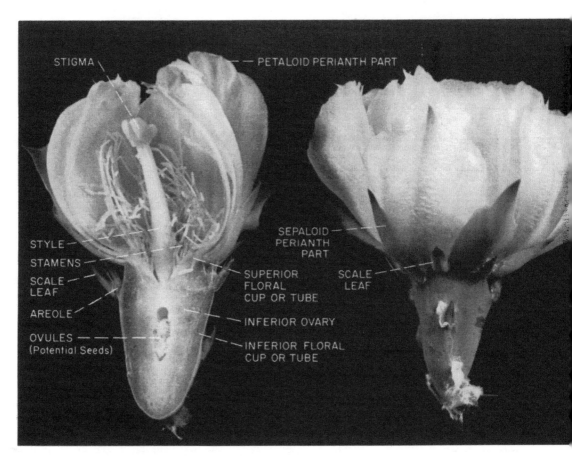

Fig. 0.3. The structure of a cactus (prickly pear) flower. In *Opuntia* (chollas and prickly pears) the floral tube is very short in its extension (superior floral tube) above the portion enveloping and adherent to the ovary wall. (Photograph by Walter S. Phillips.)

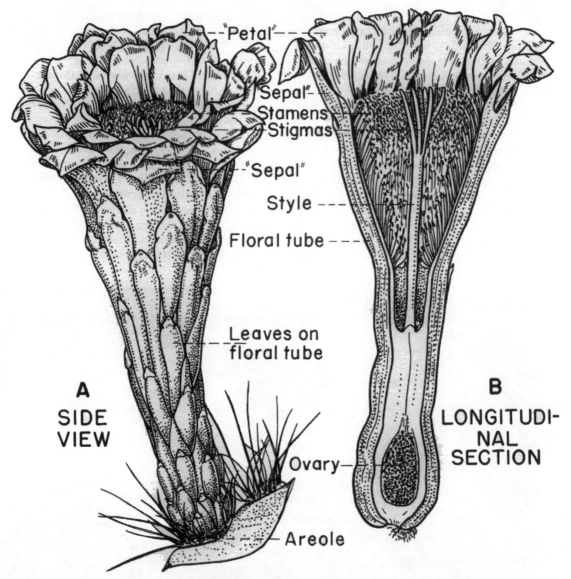

--"Petal"--

Sepal"-
Stamens
Stigmas

--"Sepal"

Style ---

Floral tube ---

Leaves on
floral tube

A
SIDE
VIEW

B
LONGITUDI-
NAL
SECTION

Ovary--

Areole

Fig. 0.4. The structure of a flower of a saguaro or giant cactus,
Cereus giganteus, a type with a well-developed superior floral tube,
the part above that enveloping and adherent to the ovary wall.

lacking as in the chollas and prickly pears. the upper part of the *pistil* continues into a single *style* with several stigmas at the apex. The *perianth* consists of several series of structures resembling sepals and petals, which intergrade with one another and with scalelike leaves (scales) on the portion of the floral tube covering and adherent to the ovary. The outermost *sepaloid* (outer and lower) *perianth parts* usually are green or brown, but sometimes they are colored. The *petaloid* (inner and upper) *perianth parts* are highly colored in most cacti, and the colors may vary between or within the species, as in other flowering plants. Numerous *stamens* are the pollen-producing organs, each consisting of a stalk or *filament* and a terminal *anther* composed of pollen sacs.

The following technical terms, in addition to those given above, are used in the keys:

Acicular, needlelike in form (and circular to broadly elliptic in cross section).

Acuminate, narrowed abruptly into a long, terminal, pointed structure.

Acute, forming an acute angle.

Adventitious root, one developed from a stem or a leaf, rather than part of the primary root system.

Annulate, with horizontal ringlike projections.

Appressed, lying flat against the stem.

Arborescent, treelike.

Arborescent plant, a woody medium-sized plant with a few stems from the base and with no single trunk, or only a very short one. Intermediate between a tree and a shrub.

Aristate (diminutive, *aristulate*), with a terminal slender point or bristle.

Caespitose, stems in a dense, low tuft.

Clublike, club-shaped, or *clavate,* elongate and increasing in diameter gradually (or near the top abruptly) upward.

Cotyledons, the embryonic leaves in the seed, these forming in the cacti the first pair of leaves conspicuous in the seedling.

Cuneate, wedge-shaped, an isosceles triangle with the attachment on the sharp point.

Deciduous, falling away.

Declined, turned downward.

Deflexed, turned downward.

Dehiscent, splitting open and releasing pollen or seeds.

Dentate (diminutive, *denticulate*), with angular teeth projecting at right angles to the margin of the organ.

Dorsoventrally, from front to back or vice versa. A structure flattened dorsoventrally has the broadest faces on the front and back, rather than the sides.

Ellipsoid, similar to elliptic, but 3-dimensional.

Elliptic, elliptical, in the form of an ellipse, *i.e.,* with both ends rounded and the length one and one half times or typically twice the diameter.

Entire, with a smooth, unindented margin.

Epidermis, the outermost layer of cells of an organ, usually secreting a layer of waxy material (cutin) which retards evaporation of water.

Fimbriate, with a fringe on the margin.

Floral tube (or, according to shape, *cup*), the covering structure (*inferior floral tube* or *cup*) forming in the cacti what appears to be the "outer coat" of the ovary and its continuation above the ovary (*superior floral tube* or *cup*) to which the perianth and the stamens are attached.

Fruit, the developed ovary and its enclosed seeds, in the cacti completely covered by and wholly adherent to the inferior floral tube (or cup), which is considered part of the fruit — an outer coat.

Globose, practically spherical.

Hilum, the scar on the seed marking the point at which the stalk (*funiculus*) of the seed was attached. This is the basal point of the seed. In seeds which are broader than long it appears to be on the side of the seed, *i.e.* to be "lateral," but this is an illusion.

Joint, a segment or section of a jointed stem.

Lanceolate, the shape of a lance, 4 to 6 times as long as broad, acute at both ends, and attached at the broader end.

Lateral, extending to the side.

Laterally, from side to side; *cf.* dorsoventrally. A structure flattened laterally has the broadest faces on the sides.

Linear, narrow, with the sides parallel and the length about eight or more times the width.

Mammillate, with the form of a human breast.

Micropyle, a pinholelike or microscopic opening in the seed-coat, this being the passageway through which the pollen tube entered the ovule.

Mucronate (diminutive, *mucronulate*), with a short terminal point of the same texture as the rest of the organ (*e.g.,* perianth part).

Obconical, conical, but attached at the point of the cone.

Oblanceolate, like lanceolate, but attached at the narrow end.

Obovate, ovate but with the attachment at the small end.

Obovoid, ovoid but with the attachment at the small end.

Obtuse, forming an obtuse angle.

Orbiculate, nearly circular.

Ovate, egg-shaped — that is, with both ends rounded — the length about one and one-half times the width, and the apical end a little narrower than the basal. Applied to objects such as leaves, which are practically two-dimensional.

Ovoid, cf. ovate. Applied to thick objects which are nearly circular in cross section — that is, distinctly three-dimensional.

Papillate, with low, rounded projections (*papillae*).

Pectinate, resembling the row of teeth in a comb.

Reticulate, forming a meshwork like "chicken-wire."

Scarious, thin and membranous and somewhat translucent, like parchment.

Serrate (diminutive, *serrulate*), sawtoothed, *i.e.,* with acutely angled, forward-projecting teeth on the margins.

Shrub, a woody, relatively small plant with several to many main stems from ground-level.

Spiniferous, spiny, spine-bearing.

Striate, with lengthwise (longitudinal) markings.

Sub-, prefix meaning almost.

Subulate, the shape of a shoemaker's awl, that is, flattened (a narrow ellipse in cross section) and tapering gradually to a terminal point.

Tree, a woody, large plant with a single main trunk that usually branches above.

Truncate, "chopped off" abruptly.

Tubercle, a projection.

Tuberculate, with tubercles.

Turbinate, with the shape of a top.

JUVENILE FORMS

Except through special study of development of juvenile forms, young cacti may be difficult to reconcile with the adult plants. For example, when *Ferocactus Wislizenii,* the common barrel cactus of southern Arizona, is 1 to 3 inches in diameter, it has none of the characteristic bristelike radial spines, which distinguish it from *Ferocactus Covillei. Coryphantha vivipara* var. *bisbeeana* and *Coryphantha Scheeri* var. *robustispina* lack central spines while the branches are young, and both old, large branches with stout, curved centrals and young branches lacking centrals may be observed on the same plant. The juvenile stem persists, of course, as the basal part of the older stem. It may or may not retain the spines. Juvenile plants of well-known species of chollas may differ so markedly from the adults in spines and tubercles as to be unrecognizable. The juvenile spines of the saguaro have been mentioned. Frequently juvenile plants have been thought to be new species and have been given formal scientific names.

Small patches of a curious cholla growing in the neighborhood of Sabino Canyon in the Santa Catalina Mountains near Tucson have remained unclassified for years. The plant, only a few inches high, advances gradually in all directions by means of rhizomes, and it has not been known to flower or to produce fruit. It may be a juvenile form that does not attain the characteristic development of an adult plant. It resembles the young forms of some of the chollas in the vicinity. There may be other juvenile vegetative forms in various localities.

Descriptions and keys given in the text of this book are based upon the characteristics of the adult plants, and they are not applicable to seedlings.

HOW TO IDENTIFY CACTI

The cacti growing in Arizona belong to eleven main groups or genera. These groups may be determined by the use of keys or of the following synopsis:

1. OPUNTIA. — Stem composed of series of cylindroid or flat joints (the cylindroid-jointed *Opuntias* or chollas [pronounced *choyas*], and the flat-jointed *Opuntias* or prickly pears, none of the other Arizona cacti except some species of *Cereus* having series of joints); areoles with glochids (sharp, barbed bristles).

2. CEREUS. — Stem with ridges and grooves on the surface; flowers produced within the spine-bearing areoles at the side of the plant or slightly below the apex of a branch; length of stem fifteen to one-hundred times the diameter.

3. ECHINOCEREUS. — Hedgehog cactus. Stem with ridges and grooves on the surface; fruit spiny; flowers produced below the stem-apex, each one just above a mature spine-bearing areole and bursting the epidermis of the stem as it grows; stem not more than eight times as long as its diameter.

4. MAMMILLARIA. — Stem with unconnected tubercles on the surface; fruit without hairs, scales, or spines; flowers produced at the side of the stem among older tubercles, growing between tubercles and having no surface connection with the spine-bearing areole.

5. FEROCACTUS. — Barrel cactus. Stem massive, with ridges and grooves on the surface; fruit fleshy, scaly, opening at the extreme base; flowers at the step-apex, each produced at the edge of the spine-bearing areole of a new portion (tubercle) of the ridge (rib).

6. ECHINOCACTUS. — Barrel cactus. Stem with ridges and grooves on the surface; fruit fleshy, then scaly, or dry, then hairy, not splitting open; flowers produced at the stem-apex; sepaloid perianth parts attenuate to spinose.

7. SCLEROCACTUS. — Stem with ridges and grooves on the surface, the upper portion of the ridge divided into tubercles; fruit dry, with or without scales and short hairs beneath them, *either* separating crosswise just above, at, or below the middle *or* splitting lengthwise along two or three lines; flowers at the stem-apex, each produced at the edge of the spine-bearing areole at the summit of a new tubercle.

8. PEDIOCACTUS. — Stem with unconnected tubercles on the surface; fruit dry, with or without a few scales, splitting open along the back and the top coming off like a lid; flowers at the stem-apex, each produced at the edge of the spine-bearing areole at the summit of a new tubercle.

9. EPITHELANTHA. — Stem with unconnected tubercles on the surface; fruit fleshy, without spines, hairs, or scales; flowers at the stem-apex, each produced at the edge of the spine-bearing areole at the summit of a new tubercle; small plants with white spines, these in an apical tuft at the growing point of the stem, the tip of each breaking away and leaving the rest of the stem with "shorter" spines.

10. NEOLLOYDIA. — Stem with the tubercles evident, not connected into ribs; fruit dry, with a few membranous scales; flowers at the stem-apex, each produced at the base of a new tubercle in a depression connected with the spine-bearing areole by a narrow groove.

11. CORYPHANTHA. — Stem with unconnected tubercles on the surface; fruit fleshy, without spines, hairs, or scales; flowers produced at the stem-apex, each at the base of the upper side of a new tubercle in a depression connected with the spine-bearing areole by a narrow groove.

Although many of the cacti may be recognized by the illustrations, or may be placed in the proper genus by the use of the preceding synopsis, precision in identification requires use of keys and descriptions. Because only a little effort is required to learn the proper method for determination of the members of a small group of plants such as the cacti, persons especially interested in these plants will be well rewarded for the effort. In the following illustration of the method of using the keys, the saguaro or giant cactus is chosen as an example because its identity is well known and unmistakable.

Turn first to the *Key to the Genera* of Arizona cacti (page 28). The lines on that page have various degrees of indentation from the left margin. The two lines or *leads* at the extreme margin (identified with the number 1) are opposed to each other. The saguaro, or giant cactus, may answer one description or the other. Reference to the section on the structure of cacti, and to the illustrations referred to there, together with examination of a living plant, will show that the saguaro has no glochids, has a leafless stem, has a floral tube about 1½ to 2 inches in length above the ovary of the flower and below the sepals and petals, has black rather than bone-white seeds, has a stem *not* made up of a regular series or "chain" of joints, and has scaly fruits that split open along three lines.

In other words, the giant cactus does not fit the description given in the upper lead of the pair, but it *is* in harmony with the lower lead 1.

The second choice in the key is between the leads identified by the number 2 and indented slightly under the lower lead 1, just chosen. The giant cactus answers the description in the upper lead 2.

The third choice is between the two leads 3 indented under the chosen lead 2. The saguaro fits the description in the first lead 3.

The fourth choice is between leads 4 indented under the chosen lead 3. The giant cactus fits the description in the first lead 4. The name of the genus *Cereus* appears at the end of the line and is preceded by the number 2, indicating that this genus is the second one discussed in the text following the key. The reader should now turn to the description of *Cereus* on page 107.

After the characteristics of the saguaro have been checked against the characteristics of the genus *Cereus,* the next step is to read through the *Key to the Species* (p. 107).

The saguaro fits the first of the two opposed leads 1 in the *Key to the Species.* Next, it fits the first lead 2 under the chosen lead 1. The full scientific name of the saguaro, *Cereus giganteus,* appears at the end of the selected lead 2. The number 1 preceding the scientific name indicates that the description and discussion of the species will precede the description of the other four species of *Cereus.*

Accurate use of the key depends upon having a full representation of the stems, spines, flowers, fruits, and seeds of the cactus. These structures are well enough illustrated here in the photographs of the saguaros to enable the reader to check some of the characteristics against the key. However, in order to identify an unknown cactus, it is necessary to have full specimens from a living plant.

PROBLEMS IN CLASSIFYING AND NAMING CACTI

The scientific name of a plant is constructed like the name of a person, but with the surname first. As stated above, the name of the giant cactus is *Cereus giganteus.* Four other kinds of *Cereus* occur in Arizona. These are *Cereus Thurberi, Cereus Schottii, Cereus* *Greggii,* and *Cereus striatus.* All five cacti of this type are thought to belong to the genus *Cereus,* and each is considered to be a separate species. (The plural of genus is genera, and the plural of species is species.)

CLASSIFICATION

Variations in acceptance of genera and species occur in different books. For example, Britton and Rose in their four-volume work on the cactus family *(The Cactaceae* 1–4: 1919–23) adopted five different generic names for the five Arizona species of *Cereus,* as follows: *Carnegiea, Lemaireocereus, Lophocereus, Peniocereus,* and *Wilcoxia.* This does not represent an error or necessarily a mistake in judgment but instead a difference of opinion concerning classification of the plants. Some botanists believe it is more advantageous to include a narrow range of forms in a genus or species; others believe it is more practical to include a broader group of variants. Each individual attempts to classify the plants of various families in a uniform manner, and there is no "right" or "wrong" system. The goal of the conservative and the liberal alike is a consistent method of classification of the plant kingdom.

The writer has presented a detailed discussion of his philosophy of genera, species, and varieties in a published report: "The Goal and Methods of Systematic Botany," *Cactus and Succulent Journal* 15:99–111, July, 1943 (cf. particularly pp. 101–103), and later in a book: *Plant Taxonomy, Methods and Principles* (Ronald Press Company, New York, 1962). In the publications mentioned, an attitude of neutrality has been maintained, because, on purely logical grouds, as long as plant populations are recognized by scientific groups that are natural, a conservative or liberal or any intermediate policy in delimiting them must be of equal value. The choice must be based, therefore, upon *(A)* conformity to prevailing practice through the world as a whole for naming the entire plant kingdom, as far as this can be determined, and *(B)* practical considerations.

A. The writer has adopted a conservative policy in all his publications, partly because it is most nearly in harmony with worldwide prevailing practice. The work of Britton and Rose represents nearly the height of a local "liberalism" endemic in the United States in the period from about 1900 to 1930. Despite the abandoning of this policy by the majority of botanists even in the United States, Britton and Rose's very useful series of volumes has not been matched by a complete comparable recent treatment of the cactus family, and there is no comprehensive coverage of the family according to conservative policy. This has resulted in retention in popular, semi-technical, and even some technical works of microgenera and microspecies of a value recognized currently by few botanists in works on other groups. To bring classification of the rest of the plant kingdom into a system comparable to that current in many books on the Cactaceae like that of Britton and Rose would require at least one million changes of plant names, and probably this is a marked understatement. To bring it into harmony with the publications of some recent authors of works on the Cactaceae would require a computer. The policy adopted in the current edition of *The Cacti of Arizona* is the approximate equivalent of that in such standard botanical books as the eight editions of Gray's *Manual of Botany* or Jepson's *Manual of the Flowering Plants of California* or *A Flora of California.*

B. Conservative policy with respect to recognition of species is followed here for the following practical reasons as well:

1. A great many fairly well-marked natural plant populations are lacking in characters of sufficient stability to make their use in keys practical. The writer believes that these populations of less stability are more effectively considered as varieties than as species. This removes the necessity of attempting to segregate them in keys, and it leaves for separation within the keys only the major populations with relatively clear and stable diagnostic characters. This saves the reader no end of frustration. When a species is represented by three or more varieties, these are differentiated in a table *(e.g., cf. Coryphantha vivipara,* p .198), thus throwing the burden of differentiation upon the complex of characters rather than upon single characters, as in a key.

2. Lack of organization of species into varieties results in consideration of each local element as a separate species, making the carrying over of knowledge from one region to another difficult. It gives the flora of each locality the aspect of being a local inde-

pendent unit rather than a phase of the general flora of the continent and of the world as a whole. If one is concerned with the flora of only a limited geographical region, the necessity for reconciling the plants of one region with those of another is not encountered, and use of a trinomial (*e.g. Prunus virginiana* var. *demissa* for the chokecherry of the Pacific States) seems cumbersome. However, for the person who has known elsewhere a closely similar plant (*e.g. Prunus virginiana,* the chokecherry of the East, Middle West, and Southeast; or *Prunus virginiana* var. *melanocarpa* of the Rocky Mountains), it is helpful to know that the plant he encounters in a new situation is similar to the one he knows already but that it does differ in some not wholly stable characters present through a particular geographical region. If the Pacific Coast chokecherry, for example, appears simply under the binomial *Prunus demissa,* there is no ready correlation with knowledge of the flora of other regions. Before the writer are half a dozen books written by members of the American "liberal" school. Each of them covers one or another of the states or larger portions of the Western States. In each of them one or two chokecherries are described, but in none of them is there any statement of close relationship to *Prunus virginiana.* In some books *Prunus demissa* appears; in some *Prunus melanocarpa* occurs; in some both are mentioned. Were these not among the relatively few plants with well-established English names, all evidence of the transcontinental relationship of the chokecherries would be obliterated. Obviously there is room for difference of opinion, but to the writer it seems far better to consider these incompletely differentiated geographical units as follows: *Prunus virginiana* var. *virginiana* (typical variety) of the East, Middle West, and Southeast; *Prunus virginiana* var. *melanocarpa* of the Rocky Mountains and the borders of the Great Basin; *Prunus virginiana* var. *demissa,* largely of the Pacific Slope but penetrating far inland in some areas. Such a policy makes "variety" a strong and useful category of value in improving organization.

In order that the generic and specific names appearing in other works may be reconciled with those used in this conservative treatment of the Arizona species, for each genus and species a list of synonyms is given in a short paragraph. Thus various points of view are made available to the reader, and he is free to choose according to his own preference.

In the introduction to the second edition of this book the following statement was made: "It is possible that a better scheme of classification than either this or Britton and Rose's may result from further study. This may lie somewhere between the very conservative viewpoint tentatively held (but with no hard and fast future commitment) by the writer and the exceedingly liberal one of Britton and Rose. However, since the interrelationships of the subgenera (or microgenera) are amazingly complex, any accurate re-evaluation and realignment will require a very long and detailed study of the entire family."

Since this statement was written, nineteen additional years of research have been applied to the cactus family, particularly to the genera and species in the United States and Canada, and a monograph of these species has been prepared. Although many problems are yet to be solved, hopefully some steps, as reflected here, have been made toward the truth. These steps are based upon many days in the field in all the states and provinces and upon visits to nearly sixty college, university, and museum herbaria to study specimens. Many decisions cannot be made or must be tentative until more data become available. As in all research, we are forever approaching the truth but we never quite reach it. In this case we are still far from it.

Although either a conservative or a liberal interpretation may be tenable on theoretical grounds, the position of either a "lumper" or a "splitter" is not. These terms connote individuals grouping as "species," "genera," or other "taxa" — groups of plants that are not natural units but mere artificial assemblages according to presence of some one or two characters to which arbitrary "importance" has been attached. Classification is not according to individual characters but to the tendency for groups of characters to be consistent in their association. This produces a natural classification system in which related plants are brought together in taxa. Any single character may or may not be in every instance a part of the association. The tendency toward "splitting" of taxa has been particularly common in treatments of the Cactaceae, and nearly always it has been associated with the arbitrary choice of a character as being so "important" as to override all other considerations. One character has been thought to be the "mark" of the taxon. Actually one character is not *per se* more important in classification than another. Its importance as a marker of a species or other category is only in proportion to its degree of consistency in association with other characters.

NAMING

Most changes in the names used for plants reflect underlying changes in classification, but some do not. Scientific naming, once classification is determined, is according to automatic application of the *International Code of Botanical Nomenclature.* The Code is based upon the following principles: (1) one taxon may have only one valid name; (2) more

than one taxon may not have the same name; (3) in cases not in agreement with the first two points the decision is based upon priority in publication of names; (4) the identity of the plant to which a name has been given is according to the type specimen designated by its author. In practice the Code and its application are much more complex. For a full explanation, *cf.* Lyman Benson, *Plant Taxonomy, Methods and Principles,* Ronald Press Co., New York, 1962.

The application of popular names to plants is far more confused than the use of scientific names, because no definite rules have been established. It is not unusual for several plants to be designated by the same English name or for a single plant to be mentioned by several popular names. In this book English, Spanish, and Indian names for cacti are included when they are available. Unfortunately the majority of Arizona cacti lack popular names, other than those such as prickly pear, which are applied to all the species in a large group.

THE GEOGRAPHICAL DISTRIBUTION OF CACTI

The cacti are practically an American group of plants. The only members of the family occurring outside of cultivation in the Old World are a few species of the genus *Rhipsalis* found in Africa and Ceylon, where they may be either native or introduced — an undetermined, much-argued point.

Cacti are characteristic of the American desert regions, but they are not confined to them. Some species occur on the ground or on the limbs of trees in the tropical rain forests, and others occur in relatively moist areas in the East and Middle West of the United States. The northern limits of the cacti include Washington, Minnesota, Wisconsin, and Massachusetts, and even the Canadian provinces of British Columbia, Alberta, and Ontario. They occur as far north as Peace River. The cactus family ranges southward to nearly all of South America.

In the United States, cacti are most abundant in Arizona and Texas and in parts of

southeastern California and New Mexico. Particularly in the desert hills and low mountain ranges in southern Arizona, the landscape is dominated by remarkably large species (*see* Frontispiece), including the saguaro or giant cactus, and the more fantastic parts of the desert vegetation are characterized by the organ pipe cactus, barrel cacti, hedgehog cacti, chollas, and prickly pears. The saguaro in particular is representative of Arizona, and, in the United States, it is practically confined to that state, although a few plants occur across the Colorado River in California. The plant sometimes becomes 50 feet tall, and it is one of the outstanding attractions of Arizona. Arizona has far more kinds of chollas than any other state. Sixty-eight cactus species with their various varieties are native in Arizona, and no other group of plants forms an equally remarkable fraction of the desert vegetation.

THE NATURAL VEGETATION OF ARIZONA

SPECIES COMPOSITION AND GEOLOGIC HISTORY OF FLORAS

In temperate North America there are nine regions of natural vegetation occupied by nine distinctive floras, which are characteristic associations of native species of plants and animals. Six of these occur in parts of Arizona. Each flora includes endemic races, varieties, species, and even genera of plants and animals. Each includes many taxa (classification units such as species) that have been associated or whose ancestral types have been together through long periods of geologic history — twenty-five to fifty million years or longer.

Usually each flora is expressed primarily in a particular ecological formation, such as grassland, desert, or forest, but not necessarily so. For example, under local ecological conditions some herbaceous plants in a forest, or their evolutionary derivatives, may be adapted to occurrence in open places within the forest. With change of conditions the open areas may become larger until they constitute considerable areas of grassland, as in the local prairies occurring in the Pacific Forest from Puget Sound to northwestern California. If this process is carried far enough, a new grassland vegetation such as the Palouse Prairie of the Pacific Northwest may evolve, but it may belong, except for invasion of some species from other sources, to the same ancestral flora as the forest.

Floras do not necessarily continue as wholly consistent units through great periods of geologic time. They are more or less modifiable in the association of some of their species. Because the ecological requirements of no two species are exactly alike, each phase of each flora has had some disassociation of its members and the taking up of new alliances with even each minor change of climate. Each vegetation type has also gained some members from time to time. For example, in the recent million years of Pleistocene time there have been four glacial periods with intervening warm periods. Even the areas not

glaciated were affected in the ice ages by lower temperatures and much greater precipitation. With the rapid fluctuation of climate all vegetation types have migrated back and forth, more or less from north to south, and vice versa. Land areas have been occupied and reoccupied by first one flora and then another, and hybrids, surviving under disturbed conditions, have given rise to new races, varieties, and species, which have replaced the ancestral ones in some, though not necessarily all, areas.

The validity of historic floras as a part of the basis for classification of vegetation is dependent upon the fact that some members of each type, or their modified descendents, do remain in association through long periods of geologic time and that although changes of climate occur in some areas they may not in others. For example, on the windward side of a relatively new mountain range, such as the Cascade-Sierra Nevada axis from southern British Columbia to Baja California, the Prevailing Westerlies, having just crossed vast expanses of warm seas, continue to deposit water on the coastal lowlands in the winter when often the land is colder than the ocean. The climate and the vegetation west of the mountains have not been affected by the mountain axis. However, as the air mass rises in crossing the new mountains, it expands, cools, and drops much rain and snow on the windward side and summits, where lowland forest species have evolved into mountain species and some more northern floral types have invaded the new areas of higher altitude. On the leeward (eastward) side of the mountains various kinds of deserts and other dry land vegetation have developed because as the air mass descends it is compressed and warmed and hence it tends to take up rather than to deposit moisture. Thus, there are deserts in the lee (in the "rainshadow") of the mountain ranges from British Columbia to Baja California. These are of different origins: (1) The desert plant communities to the north are derivatives of various temperate floral elements, especially of the oak woodland that previously covered the plain extending over the whole area west of the Rocky Mountain region from roughly the latitude of San Francisco and Salt Lake City southward to or below approximately the present Mexican boundary. Many of the plants of these deserts are related to those in the less modified oak woodlands and chaparrals (brushlands) of California to the west of the mountain axis and in southeastern Arizona beyond the range of its primary influence. (2) The desert communities to the south are derivatives primarily of elements in dry regions of northern Mexico. They are a part of the great desert systems of the southwestern United States and northern and central Mexico.

The interest of the plant taxonomist or systematic botanist is centered primarily upon the selection of species composing a vegetational unit and upon the origin and relationships of these species. These points are essential to understanding the classification of species. This, however, is true of all other considerations. The ecologist may have such primary interests as the formations and the life forms of the plants growing together, the physical and chemical factors affecting them and their association, or the interrelationships of all organisms (plant and animal) forming the community. Consequently his classification system for living vegetation may be altered in one direction or another, and his interpretation of vegetation may not be of primary importance to the taxonomist. Hopefully, in time all objectives may be met in one system of vegetation classification, but in the meantime more than one is needed.

THE NATURAL FLORAS OF ARIZONA

Each native species or variety, including those of the cacti, is associated with one or two or sometimes more of the twelve divisions of the six major floras listed below, and this is indicated in the text. Often the evolutionary development of species or varieties has paralleled the development of the vegetation types, and knowledge of this factor may be of major importance in classification. (*Cf.* Lyman Benson, Plant Classification, pp. 563–647, 1957, D.C. Health & Co., Boston; and Plant Taxonomy, Methods and Principles, 1962, The Ronald Press Co., New York.)

A. *The Mexican Desert Flora,* derived primarily from elements in the dry regions of northern Mexico and from plants developed, according to Axelrod, in pockets of rainshadow near the present deserts. Marker: the creosote bush (*Larrea divaricata*).

   1. *Arizona Desert.* The prevailing vegetation of the lowlands of central and southern Arizona. Mostly at 1,800 to 3,500 feet elevation. Marker: the saguaro (*Cereus giganteus*). The Arizona and Colorado deserts are representative in the United States of the Sonoran Deserts, markers of which are the ocotillo (*Fouquieria splendens*) and the desert ironwood (*Olneya Tesota*).

   2. *Colorado Desert.* California and Arizona near the Colorado River and Gila River, mostly below 1,500 feet elevation. Markers: The smoke tree (*Dalea spinosa*) and the California fan palm (*Washingtonia filifera*).

   3. *Mojavean Desert.* California to southwestern Utah and Mohave County, Arizona; southward to areas just south of the Bill Williams River. Mostly at 2,000 to 4,000 feet elevation. Marker: the Joshua tree (*Yucca brevifolia*).

B. *The Plains and Prairie Flora,* with relationships of many species to those occurring in Mexico. Markers: numerous herbs, especially perennial grasses.

   4. *Desert Grassland.* Southeastern Arizona, southern New Mexico, higher parts of Texas west of the Pecos River. Just above the desert below the Mogollon Rim as far northwestward as Kingman, Mohave County; well-developed here and there, as on the hillsides at Jerome, but best-developed in Santa Cruz and Cochise counties.

   5. *Great Plains Grassland.* Great Plains; high valleys in the Rocky Mountain System and on the plains and broad valleys of northern Arizona and northern New Mexico. In northwestern Arizona transitional to Desert Grassland. Mostly at 5,000 to 6,000 feet but in places up to 9,000 feet elevation. Blue gramma (*Bouteloua gracilis*) is predominant in most places.

C. *The Sierra Madrean Flora,* derived originally from an oak woodland flora that migrated northward in Oligocene and Miocene time from the area of the present Sierra Madre Occidental in northwestern Mexico and which covered in Miocene time the plains area from the latitude of San Francisco and Salt Lake City southward in the United States. Markers: oaks (*Quercus*), manzanitas (*Arctostaphylos*), madrone (*Arbutus arizonica*), junipers (*Juniperus*), pinyons (varieties of *Pinus cembroides*), sagebrush (*Artemisia tridentata*).

6. *Southwestern Oak Woodland and Chaparral.* Southeastern Arizona; southwestern New Mexico; Sonora and Chihuahua; Mexico; Texas in the Davis and Chisos Mountains, but in modified form. Occurring in Arizona at middle elevations along the Mogollon Rim and as far westward as Hualpai Mountain near Kingman. The chaparral phase occurs, for example, between Prescott and Yarnell Hill above Congress Junction. The oak woodland phase is uncommon and only slightly developed along the Mogollon Rim, but it is distinctive in Santa Cruz and Cochise counties. At 5,000 to 6,000 feet elevation in most areas. Markers: Emory oak or bellota (*Quercus Emoryi*), *Quercus oblongifolia, Quercus arizonica, Ceanothus Greggii, Arctostaphylos pungens.*

7. *Juniper-Pinyon Woodland.* Great Basin region and the plateau country of northern Arizona and New Mexico as well as the adjacent higher lands of southern Utah and southern Colorado. In Arizona mostly at 4,000 to 5,000 feet elevation. The phase in Arizona and New Mexico is characterized largely by trees and herbs of species different from those in the Great Basin. It is in an area primarily of summer rain rather than winter rain. Often occupying the areas of rocky soil on slopes where the Great Plains Grassland covers the deep soils of the valleys. Markers: junipers and pinyons.

8. *Sagebrush Desert.* Thompson River, British Columbia, to the Columbia Basin, the Great Basin, and valleys of the western Rocky Mountains. Poorly developed in northern Mohave County, Arizona, at about 4,000 to 5,000 feet elevation. Marker: sagebrush (*Artemisia tridentata*).

9. *Navajoan Desert.* The Colorado Plateau in southern Utah, southwestern Colorado, northern Arizona, and northwestern New Mexico. Mostly at 4,000 to 5,000 feet elevation. Markers: All the species of *Pediocactus* (except *Pediocactus Simpsonii*) and *Sclerocactus.*

D. *The Rocky Mountain Forest Flora,* probably present in the Rocky Mountain region for 100,000,000 years or longer. Markers: large conifers.

10. *Rocky Mountain Montane Forest.* Middle altitudes in the mountains of the Rocky Mountain system. Higher parts of the Mogollon Rim and the adjacent plateau and on the mountains rising above the plateau and the higher mountains of southeastern Arizona. In Arizona mostly at 6,500 to 8,500 feet elevation, but sometimes lower (as at Prescott) or higher (on south-facing slopes). Marker: western yellow pine (*Pinus ponderosa*), white fir (*Abies concolor*), white pine (*Pinus reflexa*).

11. *Rocky Mountain Subalpine Forest.* Rocky Mountain system in the upper portion of the forest belt. In Arizona mostly at 9,000 to 11,000 feet elevation. Markers: Engelmann spruce *(Picea Engelmannii),* Colorado blue spruce (*Picea pungens*), and alpine fir (*Abies lasiocarpa*), also corkbark fir (var. *arizonica*).

E *The Boreal Flora.* Markers: numerous perennial herbs.

12. *West American Alpine Tundra,* derivatives, mostly of Arctic species stranded on southern mountains after the four Pleistocene glaciations. Cascade-Sierra Nevada Axis; Rocky Mountain system; other mountains from Alaska to the Western States. The area above timber line. In Arizona above about 11,000 feet.

OCCURRENCE OF CACTUS SPECIES AND VARIETIES IN NATURAL VEGETATION TYPES IN ARIZONA

P   Primary occurrence
S   Secondary occurrence
s   Minor occurrence

Note: The Chihuahuan Desert is represented by some species in southeasternmost Arizona. It is not developed with complete clarity, however.

	Chihuahuan Desert	Arizona Desert	Colorado Desert	Mojavean Desert	Desert Grassland	Great Plains Grassland	Southwestern Oak Woodland and Chaparral	Juniper-Pinyon Woodland	Sagebrush Desert	Navajoan Desert	Rocky Mountain Montane Forest
1. *Opuntia Wigginsii*			P								
2. *Opuntia echinocarpa*			S	P							
3A. *Opuntia acanthocarpa* var. *coloradensis*		s	s	P							
B. Var. *major*		P	s								
C. Var. *acanthocarpa*				P							
D. Var. *Thornberi*		P		s							
4A. *Opuntia Whipplei* var. *Whipplei*						P		P	s	s	s
B. Var. *multigeniculata*				P							
5. *Opuntia spinosior*		P			P		S				s
6. *Opuntia imbricata*	P					s	P	P			
7. *Opuntia versicolor*		P									
8A. *Opuntia fulgida* var. *fulgida*		P									
B. Var. *mammillata*		P									
9. *Opuntia Bigelovii*		P	P								
10. *Opuntia leptocaulis*	P	P		s							
11. *Opuntia arbuscula*		P									
12A. *Opuntia Kleiniae* var. *tetracantha*		P			S						
13. *Opuntia ramosissima*			P								
14A. *Opuntia Stanlyi* var. *Stanlyi*		S			P						
B. Var. *Peeblesiana*		P									
C. Var. *Kunzei*		s	P								
D. Var. *Parishii*				P							
15. *Opuntia clavata*						P					
16. *Opuntia pulchella*									P		
17A. *Opuntia polyacantha* var. *rufispina*	s							P	P		
B. Var. *juniperina*							s	P	s		
C. Var. *trichophora*								P		s	
18A. *Opuntia fragillis* var. *fragilis*							s	S	P		s
B. Var. *brachyarthra*							s	P			

OCCURRENCE OF CACTUS SPECIES AND VARIETIES IN NATURAL VEGETATION TYPES IN ARIZONA

P Primary occurrence
S Secondary occurrence
s Minor occurrence

Note: The Chihuahuan Desert is represented by some species in southeasternmost Arizona. It is not developed with complete clarity, however.

	Chihuahuan Desert	Arizona Desert	Colorado Desert	Mojavean Desert	Desert Grassland	Great Plains Grassland	Southwestern Oak Woodland and Chaparral	Juniper-Pinyon Woodland	Sagebrush Desert	Navajoan Desert	Rocky Mountain Montane Forest
19A. *Opuntia erinacea* var. *erinacea*			s	P				s	S		
B. Var. *ursina*				P							
C. Var. *utahensis*								P	s		s
D. Var. *hystricina*						s		P	s		
20. *Opuntia Nicholii*										P	
21A. *Opuntia basilaris* var. *basilaris*			P	P				s	s		
B. Var. *longiareolata*				P							
C. Var. *aurea*								P	s		
D. Var. *Treleasei*				s							
22A. *Opuntia macrorhiza* var. *macrorhiza*						P		P			s
B. Var. *Pottsii*	S				P						
23A. *Opuntia littoralis* var. *Martiniana*				S				P			s
24A. *Opuntia violacea* var. *violacea*		s			P						
B. Var. *Gosseliniana*		s			P						
C. Var. *santa-rita*	s	s			P		s				
D. Var. *macrocentra*	P	s			P						
25A. *Opuntia phaeacantha* var. *phaeacantha*							S	P			s
B. Var. *laevis*		S			P		S				
C. Var. *major*	P	P		s	P	s	s	S		s	s
D. Var. *discata*	P	P	s	s	P	s	s	s			
26. *Opuntia chlorotica*		s	s	P	S			s	s		
1. *Cereus giganteus*		P	s								
2. *Cereus Thurberi*		P									
3. *Cereus Schottii*		S									
4. *Cereus Greggii* var. *transmontanus*		P									
5. *Cereus striatus*		s									

OCCURRENCE OF CACTUS SPECIES AND VARIETIES IN NATURAL VEGETATION TYPES IN ARIZONA

P Primary occurrence
S Secondary occurrence
s Minor occurrence

Note: The Chihuahuan Desert is represented by some species in southeasternmost Arizona. It is not developed with complete clarity, however.

	Chihuahuan Desert	Arizona Desert	Colorado Desert	Mojavean Desert	Desert Grassland	Great Plains Grassland	Southwestern Oak Woodland and Chaparral	Juniper-Pinyon Woodland	Sagebrush Desert	Navajoan Desert	Rocky Mountain Montane Forest
1A. *Echinocereus triglochidiatus* var. *melanacanthus*					s	s	s	P	S	P	s
B. Var. *mojavensis*					s		s	P			s
C. Var. *neomexicanus*	s							P	S		
D. Var. *arizonicus*								P			
E. Var. *gonacanthus*								P			
F. Var. *triglochidiatus*								P			
2A. *Echinocereus Fendleri* var. *Fendleri*							P	P			s
B. Var. *rectispinus*					P						
3A. *Echinocereus fasciculatus* var. *fasciculatus*		P			s						
B. Var. *Boyce-Thompsonii*		P									
C. Var. *Bonkerae*					P						
4. *Echinocereus Ledingii*					s		P				
5A. *Echinocereus Engelmannii* var. *Engelmannii*		s	P	s							
B. Var. *acicularis*		P	s	s							
C. Var. *chrysocentrus*				P							
D. Var. *variegatus*							P		P	P	
E. Var. *Nicholii*		P	S								
6A. *Echinocereus pectinatus* var. *rigidissimus*					P		s				
B. Var. *pectinatus*	P				s						
C. Var. *neomexicanus*	P				s						
1A. *Mammillaria gummifera* var. *applanata*	P				S						
B. Var. *meiacantha*	s				s	P					
C. Var. *MacDougalii*					P						
2. *Mammillaria Mainiae*		P			P						
3. *Mammillaria microcarpa*		P	s								
4. *Mammillaria Thornberi*		P									
5. *Mammillaria orestera*								P			
6A. *Mammillaria Grahamii* var. *Grahamii*		s			P						
B. Var. *Oliviae*		P			P						
7. *Mammillaria Wrightii*					S	P		S			
8. *Mammillaria tetrancistra*		s	P	s							
9. *Mammillaria lasiacantha*	P				s						

OCCURRENCE OF CACTUS SPECIES AND VARIETIES IN NATURAL VEGETATION TYPES IN ARIZONA

P   Primary occurrence
S   Secondary occurrence
s   Minor occurrence

Note: The Chihuahuan Desert is represented by some species in southeasternmost Arizona. It is not developed with complete clarity, however.

Species	Chihuahuan Desert	Arizona Desert	Colorado Desert	Mojavean Desert	Desert Grassland	Great Plains Grassland	Southwestern Oak Woodland and Chaparral	Juniper-Pinyon Woodland	Sagebrush Desert	Navajoan Desert	Rocky Mountain Montane Forest
1A. *Ferocactus acanthodes* var. *acanthodes*			P								
B. Var. *LeContei*		P	s	P							
C. Var. *Eastwoodiae*		P									
2. *Ferocactus Wislizenii*	P	P			P						
3. *Ferocactus Covillei*		P	s								
1A. *Echinocactus polycephalus* var. *polycephalus*				P							
B. Var. *xeranthemoides*									s	P	
2A. *Echinocactus horizonthalonius* var. *Nicholii*		P									
1A. *Sclerocactus Whipplei* var. *roseus*										P	
B. Var. *intermedius*								P		S	
C. Var. *Whipplei*										P	
2A. *Sclerocactus pubispinus* var. *Sileri*										P	
1. *Pediocactus Simpsonii*								S	P		
2. *Pediocactus Bradyi*										P	
3. *Pediocactus Paradinei*										P	
4. *Pediocactus Sileri*										P	
5A. *Pediocactus Peeblesianus* var. *Fickeiseniae*										P	
B. Var. *Maianus*						?					
C. Var. *Peeblesianus*										P	
6. *Pediocactus papyracanthus*						P		P			
1. *Epithelantha micromeris*	P				s						
1. *Neolloydia intertexta*	P				P						
2A. *Neolloydia erectocentra* var. *erectocentra*		s			P						
B. Var. *acunensis*		P									
3. *Neolloydia Johnsonii*				P							
1A. *Coryphantha Scheeri* var. *valida*	P				S						
B. Var. *robustispina*		s			P		s				
2A. *Coryphantha vivipara* var. *bisbeeana*					P						
B. Var. *arizonica*									P		S
C. Var. *desertii*				P							
D. Var. *Alversonii*			P	S							
E. Var. *rosea*									P		
3. *Coryphantha recurvata*					P		P				
4A. *Coryphantha strobiliformis* var. *strobiliformis*	P				S						
B. Var. *Orcuttii*					P						
5A. *Coryphantha missouriensis* var. *Marstonii*									P		s

New Taxa and Nomenclatural Recombinations

As a result of research, inevitably some new taxa come to light, and these require names. Choice of names follows rules adopted by the International Botanical Congresses and published as follows: Lanjouw, J.: 1952, 1956, 1966, International Code of Botanical Nomenclature: Regnum Vegetabile 6, 8, 46; Utrecht (text duplicated in English, French, German, and Spanish). Once a method of classification of genera, species, and varieties has been adopted, application of the rules of nomenclature is automatic. An author has no choice but to follow them. In revision of the classification of cacti, an unfortunate number of changes in nomenclature is necessary because even the taxa native in the United States have had no complete review of literature for 50 years and no thorough and critical study of their underlying relationships and classification for 110 years.

Complete consideration of the problems of classification and nomenclature is too detailed and technical for inclusion in this book. Such material is presented in the essentially completed manuscript for a larger and more inclusive and technical work, *The Cacti of the United States and Canada.* Because *The Cacti of Arizona,* edition 3, is expected to be published earlier, the International Code of Botanical Nomenclature requires valid publication of new names and new combinations for Arizona taxa to appear in this book. The coverage of each case is sufficient only for validation of the name, and discussion and elaboration should be sought in the larger book.

Valid publication of a new taxon * requires a Latin diagnosis and designation of a type specimen; recombination of an earlier name without a type specimen requires choice of a substitute (lectotype or neotype). (*Cf.* Benson, Lyman: Plant Taxonomy, Methods and Principles, Ronald Press Co., New York, 1962). Reference to the herbaria in which specimens are filed are the standard ones for botanical publication (Lanjouw, J., and F. A. Stafleu, 1952, 1954, 1956, 1959, 1964; Index Herbariorum, Eds. 1–5, International Bureau for Plant Taxonomy and Nomenclature, Utrecht.)

Opuntia Wigginsii L. Benson, **sp. nov.,** cf. p. 32. Erectuscula, caule lignoso, 3 dm. longo, ramis numerosis, articulis ovatis basi clavatis, 5-10 cm. longis, 6-9 mm. diametro, tuberculis ovatis confertis, 4.5 mm. longis, 3 mm. latis, aculeis albidis stramineo, majoribus 1 centralibus, 2-4.4 cm. longis, 0.25-0.5 mm. diametro, minoribus 5-7 undique radiantibus, floris ignotis, bacca sicca globoso-depressa, 1.2-1.9 cm. longo, 1.2-2 cm. diametro, pulvillis aculeo-latissimis confertis, seminibus regularibus, 4 mm. longo. *Type collection:* south of Quartzite, Yuma County, Arizona; 900 feet elevation, *Lyman & Evelyn L. Benson 16,465,* March 30, 1965. **Type:** *Pom 296,264.*

---

* Taxon, plural *taxa,* any botanical taxonomic category, such as the Plant Kingdom or one of its divisions, classes, orders, families, genera (plural of genus), species, or varieties.

Opuntia acanthocarpa Engelm. & Bigelow var. coloradensis L. Benson, var. nov., cf. p. 34. Caule ligneo erecto, ramis horizontalis, articulis 1.5-3 dm. longis, 2.5-3 cm. diametro, tuberculis 1.9-3 mm. latis, aculeis 10-12, 2.5-3.8 cm. longis. *Type collection:* west of South Pass, U.S. Highway 66, 23 miles west of Needles, San Bernardino County, California; Mojave Desert; 2,200 feet elevation, *Lyman Benson 10,375,* July 14, 1940. **Type:** *Pom 244,022.* **Isotype:** *Ariz 137,142.*

Opuntia acanthocarpa Engelm. & Bigelow var. major (Engelm. & Bigelow). L. Benson, comb. nov., cf. p. 35. *Opuntia echinocarpa* Engelm. & Bigelow var. *major* Engelm. & Bigelow, Proc. Amer. Acad. 3:305. 1856; in Emory, Rept. U.S. & Mex. Bound. Surv. 2: Cactaceae 56. 1859. *Type collection:* "β in Sonora." *Cf.* U.S. Senate Rept. Expl. & Surv. R. R. Route Pacific Ocean. Botany 4:49. 1857. "Mr. Schott found a stouter form farther south." "In deserts on both sides of the Colorado River, and in Sonora, Mr. *Schott.*" The collections by *Schott* have not been found. The following specimen is designated as a neotype: Headquarters of the Organ Pipe Cactus National Monument, Arizona, near the border of northwestern Sonora, *W. F. Steenbergh 5-2662-1,* May 26, 1962, *Pom 306,088.* **Neotype:** *Pom 306,088.* **Duplicate:** *Herbarium of the Organ Pipe Cactus National Monument.*

Opuntia Whipplei Engelm. & Bigelow var. multigeniculata (Clokey) L. Benson, comb. nov., cf. p. 38. *Opuntia multigeniculata* Clokey, Madroño 7:69. *pl. 4, f. A.* 1943. *Type collection:* Wilson's Ranch, Charleston Mountains, Nevada. *Clokey 8430.* **Type:** not found at *UC* in 1965. **Isotypes:** *UC 872,689* (box), *Pom 265,219, 275,347, Mo 1,244,341, Ill, BH, NY, WillU 30,290, 30,292, OSC, GH, US 1,828,523, BM, SMU, DS 270,338* (box & photos), *Ariz 21,473, UO, Tex, Mich, F, Ph 815,140. UC 872,689* is designated as a lectotype for so long as the type is missing. **Neotype:** *UC 872,689.*

Opuntia Stanlyi Engelm. var. Peeble-siana L. Benson, var. nov., cf. p. 64. Articulis elongatis clavatis, 7.5 cm. longis, 1.5-2.5 cm. latis, tuberculis segregis, 1.5-2.5 cm. longis, petalis flavis, bacca oblonga spinulosa, seminibus magnis. *Type collection:* 5 miles southeast of Alamo (old settlement at the crossing of the Bill Williams River), Yuma County, Arizona, 1,500 feet elevation; base of the Harcuvar Mountains; Colorado Desert; sandy flat, *Lyman Benson 10,064,* March 27, 1940. **Type:** *Pom 274,040.*

Opuntia polyacantha Haw. var. rufispina (Engelm. & Bigelow) L. Benson, comb. nov., cf. p. 70. *Opuntia missouriensis* DC. var. *rufispina* Engelm. & Bigelow, Proc. Amer. Acad. 3:300. 1856; U.S. Senate Rept. Expl. & Surv. R. R. Route Pacific Ocean. Botany 4:45. *pl. 14. f. 1-3.* 1857. *Type collection:* The following specimen is designated as a lectotype: "Rocky places, Pecos," *J. M. Bigelow,* Sept. 24, 1853. **Lectotype:** *Mo.*

Opuntia polyacantha Haw. var. juniperina (Britton & Rose) L. Benson, comb. nov., cf. p. 72. *Opuntia juniperina* Britton & Rose, *Cactaceae* 1:197. *f. 243-244.* 1919. *Type collection:* Cedar Hill, San Juan County, New Mexico, *Paul C. Standley 8051.* **Type:** *US 686,991.* **Isotype:** NY (marked "co-type").

Opuntia erinacea Engelm. & Bigelow var. utahensis (Engelm.) L. Benson, comb. nov., cf. p. 78. *Opuntia sphaerocarpa* Engelm. & Bigelow var. *utahensis* Engelm. Trans. St. Louis Acad. Sci. 11:199. 1863, not *Opuntia utahensis* J. A. Purpus, Monatschr. Kakteenk. 19:133. 1909. *Type collection:* The following specimen is designated as a lectotype: Pass west of Steptoe Valley, Nevada, *Henry Engelmann,* July 19, 1859. **Lectotype:** *Mo.*

Opuntia macrorhiza Engelm. var. Pottsii (Salm-Dyck) L. Benson, comb. nov., cf. p. 89. *Opuntia Pottsii* Salm-Dyck, Cact. Hort. Dyck. 1849: 236. 1849. *Type collection:* Chihuahua, *Potts.* No type specimen has been found. The following specimen is designated as a neotype: "Vicinity of Chihuahua; altitude 1,300 meters," *Edward Palmer 124,* April 8-27, 1908. **Neotype:** *Mo. 1,797,126.* **Duplicate:** *US.*

Opuntia violacea Engelmann in Emory, Notes Mil. Reconn. Ft. Leavenworth to San Diego. App. 2. 156. *f. 8.* Feb. 17, 1848. The identity of this species has been established by visiting the route and the campsites of the military reconnaissance expedition. The following specially collected topotype is designated as a neotype: northeast of Solomon, *Lyman Benson 16,632,* April 22, 1966. **Neotype:** *Pom 311,337.* A full discussion is included in the larger manuscript mentioned above.

Opuntia violacea Engelm. var. **Gosseliniana** (Weber) L. Benson, **comb. nov.,** cf. p. 92. *Opuntia Gosseliniana* Weber, Bull. Soc. Acclim. France 49:83. 1902. *Type collection:* Coast of Sonora, *Leon Diguet.* **Type:** *P* (two flowers).

Opuntia violacea Engelm. var. **santa-rita** (Griffiths & Hare) L. Benson, **comb. nov.,** cf. p. 92. *Opuntia chlorotica* Engelm. var. *santa-rita* Griffiths & Hare, N. Mex. Agric. Exp. Sta. Bull. (60): 64. 1906. *Type collection:* Santa Rita Mountains, Arizona, *David Griffiths 8157.* **Type:** *US.* **Isotype:** *Pom 287,241.*

Opuntia violacea Engelm. var. **macrocentra** (Engelm.) L. Benson, **comb. nov.,** cf. p. 92. *Opuntia macrocentra* Engelm. Proc. Amer. Acad. 3:292. 1856; in Emory, Rept. U.S. & Mex. Bound. Surv. 2: Cactaceae 49. *pl. 75. f. 8* (seeds). 1859. Type collection: The following specimen is designated as a lectotype: "Sandhills in the Rio Grande bottom, near El Paso," *Charles Wright* in 1852 (May, 1852, on one sheet). **Lectotype:** *Mo.* (on 2 sheets).

Opuntia phaeacantha Engelm. var. **laevis** (Coulter) L. Benson, **comb. nov.,** cf. p. 98. *Opuntia laevis* Coulter, Contr. U.S. Nat. Herb. 3:419. 1896. *Type collection:* "Type *Pringle* of 1881 (distributed as *O. angustata*) in Herb. Coulter. Arizona." **Type:** *F. 98,070* **Isotypes:** *Mo, K, Vt, GH, US 795,860.*

Echinocereus triglochidiatus Engelm. var. **arizonicus** (Rose ex Orcutt) L. Benson, comb. nov., cf. p. 129. *Echinocereus arizonicus* Rose ex Orcutt, Cactography 3. 1926. *Type collection:* Superior-Miami Highway, *Charles Russell Orcutt,* July, 1922. The specimen in the New York Botanical Garden is designated as a lectotype. **Lectotype:** *NY* (photographs, *US, Pom 313,363*). **Duplicate:** *US 73,376.*

Echinocereus fasciculatus (Engelm.) L. Benson, **comb. nov.,** cf. p. 132. *Mammillaria fasciculata* Engelm. in Emory, Notes Mil. Reconn. Ft. Leavenworth to San Diego. App. 2. 156. *f. 2.* Feb. 17, 1848. *Type collection:* Complete discussion of this complex nomenclatural problem is included in the larger manuscript mentioned above. In the absence of a type specimen, one specially collected at the type locality is designated as a neotype: south side of the Gila River near the mouth of Bonanza Creek; 3,100 feet elevation, *Lyman Benson 16,633,* April 22, 1966. **Neotype:** *Pom 311,339.* This is one of several collections made in a survey by the writer of the route and the campsites of the military reconnaissance of 1846 under command of Lieutenant W. H. Emory. No *Mammillaria* or *Coryphantha* growing in the area could be the basis for the illustration or the description. The plant was clearly the one described in the second edition of the *Cacti of Arizona* as *Echinocereus Fendleri* var. *robustus.* The plant long known mistakenly as *Mammillaria fasciculata* occurs only far to the westward on the Papago Indian Reservation south of the lower Gila. The earliest name actually applied to it is *Mammillaria Thornberi.*

Echinocereus fasciculatus Engelm. var. **Boyce-Thompsonii** (Orcutt) L. Benson, **comb. nov.,** cf. p. 133. *Echinocereus Boyce-Thompsonii* Orcutt, Cactography 4. 1926. *Type collection:* Boyce Thompson Southwestern Arboretum, Superior, Arizona. No specimen preserved. The following specimen is designated as a neotype: "56 miles east of Phoenix; 9 miles west of Superior," about 5 miles west of the Boyce-Thompson Arboretum, 2,400 feet elevation, *Lyman Benson 14,621,* October 17, 1950. **Neotype:** *Pom 278,845.*

Echinocereus fasciculatus (Englem.) L. Benson var. **Bonkerae** (Thornber & Bonker) L. Benson, **comb. nov.**, cf. p. 136. *Echinocereus Bonkerae* Thornber & Bonker, Fantastic Clan 71-73, 85. *pl. opposite 28, 72.* 1932. *Type collection:* none given. "We are nearing the beautiful Pinal Mountains in southeastern Arizona . . ." According to A. A. Nichol, the plant described by Professor Thornber was from near Oracle. Mr. Nichol planted the specimens in the University of Arizona Cactus Garden, and in 1940 the writer placed the remains of the last individual in the University Herbarium. This specimen, with a sheet of flowers from Thornber and Bonker (San Carlos, Gila County), is now designated as a lectotype. **Lectotype:** *Ariz 156,240* (box) and (for flowers) sheet *156,240.*

Echinocereus Engelmannii Parry var. **acicularis** L. Benson, **var. nov.**, cf. p. 138. Cylindricus, e basi 5-50 ramosus, 1.5-2 dm. longis, 3.8-5 cm. diametro; aculeis centralibus fiavis vel rubellis rectis gracilis, inferioribus 1 2.5-3.8 cm. longis, 1 mm. latis, superioribus 3 1.9-2.5 cm. diametro, floribus circa 6.2 cm. diametro. *Type collection:* Crossing of New River, south side of Black Canyon Refuge, Maricopa County, Arizona; 1,300 feet elevation, *Lyman Benson 16,616,* April 20, 1966. **Type:** *Pom 311,313.* This is the common variety in central Arizona.

Mammillaria gummifera Engelm. in Wisliz. Mem. Tour. No. Mex. 106 (22). *Jan. 13,* 1848. *Type collection:* "Cosiquiriachi, Dr. Wislizenus. Oct. 1848." **Type:** *Mo.* The identity of this species is established by the type material, studied during the last several years.

Mammillaria gummifera Engelm. var. **applanata** (Engelm.) L. Benson, **comb. nov.**, cf. p. 150. *Mammillaria applanata* Engelm. in Wisliz. Mem. Tour. No. Mex. 105. Jan. 13, 1848 (not described individually); Pl. Lindh. II., Bost. Jour. Nat. Hist. 6:198. 1850. *Type collection:* The following specimen is designated as a lectotype: "On the Pierdenales [Perdenales River], western [central] Texas," *F. Lindheimer* in 1845. **Lectotype:** *Mo.* A metanym is as follows: *Mammillaria Heyderi* Mühlenpfordt, Allg. Gartenz. 16:20. *Jan. 15,* 1848.

Mammillaria gummifera Engelm var. **meiacantha** (Engelm.) L. Benson, **comb. nov.**, cf. p. 151. *Mammillaria meiacantha* Engelm. Proc. Amer. Acad. 3:263. 1856: U.S. Senate Rept. Expl. & Surv. R. R. Route Pacific Ocean. Botany 4:27. 1857; in Emory, Rept. U.S. & Mex. Bound. Surv. 2: Cactaceae 9. *pl. 9. f. 1-3.* 1859. The following specimen is designated as a lectotype: "Cedar plains east of the Pecos," *J. M. Bigelow,* Sept. 27, 1853. **Lectotype:** *Mo.*

Mammillaria gummifera Engelm. var. **MacDougalii** (Rose) L. Benson, **comb. nov.**, cf. p. 151. *Mammillaria MacDougalii* Rose in Bailey, Stand Cyclop, Hort. 4:982. 1916. *Type collection:* Santa Catalina Mountains, Arizona, *D. T. MacDougal,* Nov. 1909. **Lectotype:** *US 1,821,109* (one small juvenile plant and material in two cartons). **Duplicate:** *Pom 306,413.*

Mammillaria orestera L. Benson, **sp. nov.**, cf. p. 155. Globosa seu demum ovata, simplex seu e basi ramosa, tuberculis ovato-cylindricis, 6-12 mm. longis, 4.5 mm. latis, areolis orbiculatis, aculeo centrali singulo longiore sursum hamato, 9-12 mm. longo, aculeis radialibus uniseriatis 14-24 apice fuscatis basi bulbosis, 8-10.5 mm. longis, sepalis lanceolatis ciliolatis, 6-9 mm. longis, 2 mm. latis, petalis sub-10 lanceolatis integris vel paucidentatis viridis vel roseatis, 9 mm. longis, 2 mm. latis, stigmatibus 3-5 viridis, 1 mm. longis, 0.3-0.4 mm. latis, bacca ovato-globosa succosa purpurascente, seminibus obovatis basi acutis nigris vel fusco-atris scrobiculatis, 1-1.5 mm. longis, 0.7-1 mm. latis. *Type collection:* Santa Catalina Mountains, Pima County, Arizona, *Evelyn, Lyman, & Robert L. Benson 14,864,* April 9, 1952. **Type:** *Pom 285,759.*

Mammillaria Grahamii Engelm. var. **Oliviae** (Orcutt) L. Benson, **comb. nov.**, cf. p. 161. *Mammillaria Oliviae* Orcutt, West. Amer. Sci. 12:163. 1902. *Type collection:* The following specimen is designated as a lectotype: near Vail, Pima County, Arizona, *Charles Russell Orcutt* in 1921. **Lectotype:** *US 1,821,092* and the material in the carton. **Duplicate:** *Pom 306,414.*

Ferocactus acanthodes Lemaire var. **Eastwoodiae** L. Benson, var. nov., cf. p. 166. Cylindricus, aculeis flavis vel stramineis, aculeis centralibus 7.5-8 cm. longis, circa 2.5 mm. latis, robustis compressis annulatis 4-angulatis cruciatis, inferiore deorsum curvato, aculeis radialibus 12-14, 4.4-5.6 cm. longis, 0.5-1 diametro, rectis rigidis. *Type collection:* Arizona in Pinal County near Queen Creek; rocky canyonside at 3,200 feet, *Lyman Benson 16,618,* April 20, 1966, fruits and flowers added July 1, 1966. **Type:** *Pom 311,312.*

Echinocactus horizonthalonius Lemaire var. **Nicholii** L. Benson, var. nov., cf. p. 175. Depresso-globosus vel ovatus vel cylindricus, 0.5-4.5 dm. longo, 1.25-1.5 vel 2 dm. diametro, aculeis 8 robustis compressis annulatis rubellis demum cinereis vel nigris, aculeis centralibus 3, inferiore deorsum curvato, superiore latiore sursum suberecto curvato, aculeis radialibus 3 curvato. *Type collection:* Arizona in Pima County, southwest of Silver Bell, Silver Bell Mountains, 2,800 feet elevation, Arizona Desert, *Lyman Benson 16,663,* July 3, 1966. **Type:** *Pom 311,314.*

Sclerocactus pubispinus Engelm. var. **Sileri** L. Benson, var. nov., cf. p. 179. Aculeis centralibus compressus albidis demum fusco-atris, summo complanato recto albido, ceteris plerumque breviore compresso-quadrangulatis, inferiore deorsum hamato. *Type collection:* "Southern Utah," *A. L. Siler* in 1888. **Type:** *Ph.* **Isotype:** *US* (fragments, sheet unnumbered).

Pediocactus Peeblesianus (Croizat) L. Benson, Cactus & Succ. Jour. 34:59-61. *f. 41.* 1962. The paper cited above includes a full discussion of the nomenclature of a variety of this species, according to the information available in 1962. It includes provisional acceptance of valid publication of the name combination, *Navojoa Fickeisenii* Backeberg, Cactus & Succ. Jour. Gt. Brit. 22:49 (*2 unnumbered figs. on p. 54).* 1960, nom. nud. (for lack of a type specimen); Cactaceae 5:2875. *f. 2,700-2702.* 1961 (in 1962 presumed valid because of the following statement: "Der Typus befindet sich in meiner Sammlung"). Backeberg (personal communication) did

not accept the value of plant specimens required under the International Code of Botanical Nomenclature and used for centuries by botanists, insisting that photographs were better, and it was pointed out (personal communication) that the best answer is both. Mr. Paul C. Hutchison, who talked with the late Mr. Backeberg at his home in Hamburg, Germany, reports (personal communications) that there is no collection of pressed specimens and that there are few living plants, Backeberg having depended upon the plants of vague sources growing in gardens and those sent to him. The bulk of his voluminous work was developed through amazingly astute use of information published by others, in conjunction with seeing plants in cultivation and with some early travel in Latin America. It is without documentation.

Thus, there is no evidence or likelihood of a type specimen at the time of publication,[*] and, unless preservation of a type in or before 1961 can be proved, *Navajoa Fickeisenii* is a *nomen nudum.* Consequently, all the recombinations of the epithet also are *nomina nuda* as follows: *Pediocactus Peeblesianus* (Croizat) L. Benson var. *Fickeisenii* L. Benson, Cactus & Succ. Jour. 34:59. 1962; *Toumeya Fickeisenii* W. H. Earle, Cacti Southw. 98. Feb., 1963; *Pediocactus Fickeisenii* L. Benson ex W. H. Earle, *ibid., pro. syn.,* inadvertently ascribed to the writer; *Toumeya Fickeisenii* Kladiwa, Sukkulentenkunde 7/8:46. March, 1963. Thus, the epithet *Fickeisenii* has not been published validly, and it has no status in taxonomic nomenclature. Another epithet could be substituted,

---

[*] According to Hutchison, letter of June 19, 1967, "... when I visited Backeberg in Hamburg in 1960 he told me that he never had and never would make a herbarium specimen of a cactus.... in 1960 there definitely was no Backeberg herbarium, and he had no plans to start one. He moved closer to the type concept only by agreeing 'in future publications' to cite the collection in which he had seen the Clonotype from which his description ... [was] drawn. His yearly travels throughout Europe provided the material — all in private [living] collections and practically nothing at Hamburg except a back porch with a few cacti on it." The useful term clonotype has no official status, and a living ephemeral clonotype is not the required permanently preserved type specimen necessary in botanical nomenclature.

but this would be unwise because this one has become well known through its use in various combinations. The best course seems to be to give it valid publication as new, to correct the form of the word, and to anchor it securely to a type specimen so that its identity will be established for all time. This does not validate any of the name combinations above.

This plant was discovered by Miss Maia Cowper of Belen, New Mexico, in 1956; later it was found independently at some distance by Mrs. Florence Fickeisen of Phoenix, who sent plants to Backeberg. Obviously, both individuals cannot be commemorated in naming this plant. The plants collected by the Cowpers on the Little Colorado River differ somewhat from those collected by Mrs. Fickeisen northwest of the Grand Canyon, but the remarkable variation in collections and especially in plants collected and observed near the Little Colorado by the writer in 1957, 1959, and 1961, indicates the presence of only one taxon among the known populations along the Colorado and the Little Colorado rivers (*ibid., f. 41; f. 8.4* of this book).

However, another unnamed variety of the same species has come to light through an old specimen at the United States National Herbarium, and this is named for Miss Cowper, as var. *Maianus,* cf. below.

Pedioactus Peeblesianus (Croizat) L. Benson var. **Fickeiseniae** L. Benson, **var. nov.,** cf. p. 186. Globosa vel obovata simplex seu e basi ramosa, 2.5-3.8 vel 6.2 cm. longus, 2.5-3.8 cm. diametro, aculeo centrali singulo robustiore longiore vel breviore, aculeis radialibus 6-7, 3-6 mm. longis, 0.25-0.5 mm. diametro. *Type collection:* Watershed of the Little Colorado River west of Cameron, Coconino County, Arizona, the following two collections considered as one, both having been made in exactly the same place, one at fruiting, the other at flowering time: *Lyman & Robert L. Benson 15,745,* June 28, 1957, *Pom 285,856* and *Lyman and Evelyn L. Benson 16,086,* April 21, 1961. *Pom 299,969.* **Type:** *Pom 285,856 & 299,969.*

Pediocactus Peeblesianus (Croizat) L. Benson var. **Maianus** L. Benson, **var. nov.,** cf.

p. 186. Ovato-cylindricus, simplex, 6.2 cm. longus, 3.8 cm. diametro, aculeis centralibus 0, aculeis radialibus 6, 1.2 cm. longis, 3 superioribus debilioribus, lateralibus brevioribus, 3 inferioribus robustioribus infra curvato. *Type collection:* Prescott, Arizona, *J. W. Toumey,* April 23, 1897. **Type:** *US 535,244.*

The following epithet, not validly published, was intended for a form of *Pediocactus Peeblesianus* var. *Fickeiseniae* (*not* var. *Maiana,* as published here) from the lower Little Colorado River. *Navajoa Maia* Cowper, Cactus & Succ. Jour. Gt. Brit. 23:90. 1961, *nom. nud.;* 24:16 *(photograph).* 1962. *nom. nud.* This name, appearing in journals and catalogs, is mentioned to avoid confusion.

Neolloydia intertexta (Engelm.) L. Benson, **comb. nov.,** cf. p. 191. *Echinocactus intertextus* Engelm. Proc. Amer. Acad. 3:277. 1856; in Emory, Rept. U.S. & Mex. Bound. Surv. 2: Cactaceae 27. *pl. 35.* 1859, not Philipi, Linnaea 33:81. 1864. *Type collection:* "From El Paso to the Limpio [Limpia Creek], and southward to Chihuahua. . . ." "On stony ridges from the Limpia to El Paso, Wright, Bigelow, and westward, Parry; also toward Chihuahua, Wislizenus. . . ." The specimen from El Paso in 1852 (mounted on two herbarium sheets) is designated as a lectotype. There is no indication of whether this specimen was collected by Wright or Bigelow. Part was from cultivation in St. Louis in 1855. **Lectotype:** *Mo.* **Duplicate:** *Pom.*

Neolloydia erectocentra (Coulter) L. Benson, **comb. nov.,** cf. p. 191. *Echinocactus erectocentrus* Coulter, Contr. U.S. Nat. Herb. 3:376. 1896. *Type collection:* "Type in Nat. Herb. and in Herb. Coulter. Near Benson, Arizona, and also near Saltillo, Coahuila. Specimens examined: ARIZONA (*Evans* of 1891:) COAHUILA (*Weber* of 1869)." The specimens have not been found, and the following collection is designated as a neotype: east of the junction of U.S. 80 and the road to Sonoita, Pima County, Arizona, *L. Benson 10,326,* April 17, 1940, *Pom 273,980.* A fruiting specimen from the same place is *L. Benson 16,673,* July 15, 1967, *Pom 311,-315.* **Neotype:** *Pom 273,980.*

Neolloydia erectocentra (Coulter) L. Benson var. **acunensis** L. Benson **comb. nov.,** cf. p. 192. *Echinocactus acunensis* W. T. Marshall, Saguaroland Bulletin 7:33. *photograph.* 1953. *Type collection:* The following specimen is designated as a lectotype: "Acuna Valley, Organ Pipe Natl. Mon., Pima Co., Arizona, Altitude 1300 to 1850 ft.; rocky hills; creosote bush cover," Wm. Supernaugh, Jan. 2, 1951. **Lectotype:** *Des.* **Duplicate:** *Herbarium of the Organ Pipe Cactus National Monument.*

Neolloydia Johnsonii (Parry) L. Benson, **comb. nov.,** cf. p. 192. *Echinocactus Johnsonii* Parry ex Engelm. in S. Watson in King, U.S. Geol. Expl. 40th. Par. Botany 117. 1871. *Type collection:* The following specimen is designated as a lectotype: "Discovered about St. George in Southern Utah by *J. E. Johnson* in 1870." **Lectotype:** *Mo.*

Coryphantha Scheeri (Kuntze) L. Benson, **comb. nov.,** cf. p. 195. *Mammillaria Scheeri* Mühlenpfordt, Allg. Gartenz. 15:97. 1847, not *Mammillaria Scheeri* Mühlenpfordt, Allg. Gartenz. 13:346. 1845. *Coryphantha Scheeri* Lemaire, Cactées 35. 1868, *nom. nud.,* perhaps intended for this species. The species of *Coryphantha* listed by Lemaire were based upon those of *Mammillaria* bearing the same epithet, but in this case there are two earlier versions of *Mammillaria Scheeri,* and the reference is uncertain. *Cactus Scheeri* Kuntze, Rev. Gen. et. Sp. *Pl.* 1:261. 1891, the first valid publication of the epithet, it being considered as newly published in 1891 (International Code of Botanical Nomenclature, Article 72). *Type collection:* "Original Exemplare . . . in Sammlung des Herrn Scheer zu Kew, des Herrn Fürsten zu Salm-Dyck, und in meiner." Existence of a type specimen or a possible lectotype has not yet been determined.

Coryphantha Scheeri (Kuntze) L. Benson var. **valida** (Engelm.) L. Benson, **comb. nov.,** cf. p. 195. *Mammillaria Scheeri* Mühlenpfordt var. *valida* Engelm. Proc. Amer. Acad. 3:265. 1856; Rept. U.S. & Mex. Bound. Surv. 2: Cactaceae 10. 1859, not *Mammillaria valida* J. A. Purpus, Monatschr. Kakteenk. 21:97. 1911 (which probably is

var. *Scheeri* from Mexico). *Type collection:* "Sandy ridges of the Rio Grande near El Paso: fl. July." "Sandy ridges of the valley of the Rio Grande from El Paso to the Cañon; also at Eagle Springs and on prairies at the head of the Limpia:" Charles Wright. The following specimen is designated as a lectotype: "Prairies at the head of the Limpio," *Charles Wright,* June 25, 1852. **Lectotype:** *Mo.*

Coryphantha Scheeri (Kuntze) L. Benson var. **robustispina** (Schott) L. Benson, **comb. nov.,** cf. p. 195. *Mammillaria robustispina* Schott ex Engelm. Proc. Amer. Acad. 3:265. 1856; Rept. U.S. & Mex. Bound. Surv. 2: Cactaceae 11. *pl. 74. f. 8.* 1859. *Type collection:* "On grassy slopes on the south side of the Babuquibari [Baboquivari] mountains, in Sonora; A. Schott. Flowers July." **Type:** *Mo.* **Isotype:** *F 42,679* (Schott Herbarium).

Coryphantha vivipara (Nutt.) Britton & Rose var. **bisbeeana** (Orcutt) L. Benson, **comb. nov.,** cf. p. 197. *Coryphantha bisbeeana* Orcutt, Cactography 3. 1926. *Type collection:* "See note under C. aggregata," as follows: "This is based upon Engelmann's Mammillaria aggregata, which, according to that author was Echinocereus phoeniceus (E. aggregata Coulter [Rydberg]). B. & R. 4: t. 4. [Britton & Rose, Cactaceae 3: *pl. 4,* 1922] represents C. Arizonica; f. 47 represents C. Bisbeeana Orc." Orcutt's intention was clearly to rename the plant described as *Coryphantha aggregata* by Britton & Rose (Cactaceae 3:47. *f. 47.* 1922), as *Coryphantha bisbeeana.* This was because he adopted Engelmann's reinterpretation of *Mammillaria aggregata* (in Ives, Rept. Colo. R. West., part 4, *Botany* 13. 1861), which is not mentioned by Britton & Rose and which seems to have been overlooked by all other authors. The only collection by Rose (U.S. National Herbarium staff) from Benson, Arizona, is *Rose 11,958,* the probable basis for Britton & Rose's figure 47. The specimen is designated as a lectotype for *Coryphantha bisbeeana* Orcutt, which was based upon *Coryphantha aggregata* as interpreted and described by Britton & Rose, not upon the basionym of *Mammillaria aggregata* Engelm. **Lectotype:** *US* (box).

Coryphantha vivipara (Nutt.) Britton & Rose var. **Alversonii** (Coulter) L. Benson, **comb. nov.**, cf. p. 200. *Cactus radiosus* (Engelm.) Coulter var. *Alversonii* Coulter, *Contr. U.S. Nat. Herb.* 3:122. 1894. *Type collection:* The following specimen is designated as a lectotype: "Mojave Desert, Calif., A. H. Alverson. Mr. Alverson's specimens were collected at McHaney's Mine near 29 Palms. S. B. P. [Parish]." **Lectotype:** *UC 205,017.*

Coryphantha vivipara (Nutt.) Britton & Rose var. **rosea** (Clokey) L. Benson, **comb. nov.**, cf. p. 200. *Coryphantha rosea* Clokey, *Madroño* 7:75. 1943. *Type collection:* Charleston Mountains, Clark County, Nevada, *Ira W. Clokey 8,038*, June 24, 1938. **Type:** *UC 905,407.* **Isotypes:** *Pom 265,218, 275,-343, Mo* (including a very large plant) *NY (2), OSC, UO, GH, Ill, US 1,828,522, Ph 815,138, BH* (box), *Mich, F, UC 1,102,655, SMU, Ariz 47,693.* The type includes a stem 5 inches long and 4 in diameter; most of the isotypes are from smaller, presumably younger, plants.

Coryphantha strobiliformis (Poselger) Orcutt var. **Orcuttii** (Rose ex Orcutt) L. Benson, **comb. nov.**, cf. p. 204. *Neolloydia Orcuttii* Rose ex Orcutt, *Cactography* 5. 1926, *nom. nud.;* in Möller's *Deutsch. Gärtnerzeitung* 142. 1926. *Escobaria Orcuttii* Bödeker, *Mam.-Vergl.-Schlüssel* 17. 1933. *Type collection:* "NM [New Mexico]." The following specimen is designated as a lectotype: *"Escobaria Orcutti* Rose, ined. Near Granite Pass, N.M. March 1926 (type locality) C. R. Orcutt." **Lectotype:** *DS 307,410* (box).

Coryphantha missouriensis (Sweet) Britton & Rose var. **Marstonii** (Clover) L. Benson, **comb. nov.**, cf. p. 204. *Coryphantha Marstonii* Clover, *Bull. Torrey Club* 65:412, *pl. 17. f. 6,* 1938. "Specimen typicum ([no.] 1909) vivum ex loco 'Boulder' dicto conservatum est in Horto Botanico Universitatis Michiganensis." "Known only from the type locality, 'Hell's Backbone,' a mountain ridge near Boulder, Garfield Co., Utah ([no.] 1909)." No type specimen was preserved. The following specimen is designated as a neotype: Utah; Kane County; east side of Buckskin Mountains, 5,200 feet elevation, *Lyman Benson 15,205,* August 8, 1953. **Neotype:** *Pom 285,320, 296,309.*

Classification of Cacti

ORDER *CACTALES:* CACTUS ORDER

FAMILY *CACTACEAE:* CACTUS FAMILY
(The only family)

Stems simple or branching, succulent, the larger ones minute to 50 feet long and as much as 2½ feet in diameter; ribs none to many, the tubercles separate or coalescent or partly so. Leaves when present ranging from persistent, large, and flat to ephemeral, small, and conical to cylindroidal, but usually not discernible in the mature plant. Areoles produced fundamentally in the leaf-axils, bearing spines at least in the juvenile stages of the stem, these variable. Flowers and fruits on either the new or the old growth of the current or preceding seasons, accordingly near or below the apex of the stem or branch, each either at the apex or on the upper side of a tubercle or between the tubercles, either near or distant from the spine-bearing part of the areole, usually either merging with it or connected by an isthmus, this area persisting for many years after the fall of the fruit and leaving a circular, irregular, oblong, or elongate and often narrow scar. Flower epigynous or in *Pereskia* perigynous; floral tube a hypanthium, either bearing areoles and usually also small leaves, or bare, above its junction with the ovary ranging from almost obsolete to elongate and tubular; leaves or scale-leaves of the floral tube covering the ovary shading into the sepaloid outer perianth parts and these into the petaloid (usually highly colored) inner parts; stamens numerous, arranged spirally; carpels 3 to more than 20; stigmas separate; style 1; ovary with a single chamber and numerous ovules on (as seen at maturity) a parietal placenta (on the ovary wall, but the placentae actually complex); fruit fleshy or dry at maturity, with tubercles, scales, spines, hairs, or glochids or without surface appendages, of various shapes and sizes. Seeds black, brown, white, reddish, gray, or bone-white, of various shapes, ranging from longer than broad to broader than long (length being hilum to opposite side), mostly 1/24 to 1/5 inch in greatest dimension; hilum either obviously basal or appearing "lateral"; embryo with the cotyledons lying either parallel or at right angles to the flat faces of the seed as a whole and with either the edges of both cotyledons or the back of only one cotyledon toward the hypocotyl, therefore either accumbent or incumbent or sometimes oblique.

DISTRIBUTION: An undetermined number (perhaps 800 or 1,000 or more) of *valid* species occurring from Canada to southern South America; 147 species and 127 varieties other than the nomenclaturally typical ones (actually beyond one) occurring as native, or in a few cases introduced plants in Canada and the fifty states of the United States; 68 of the species and 55 of the varieties occurring in Arizona.

27

KEY TO THE GENERA

1. Glochids (small or minute, sharp-pointed, barbed bristles) as well as longer and stouter spines developed in the areoles; young stem with a fleshy leaf at the base of each areole; floral tube *not* extending above the ovary; seeds bony, often discoid; stems composed of series of cylindroid or flattened joints; flowers and fruits occurring on the joints produced the preceding year; flower produced within or on the edge of the areole, all of which is spine-bearing; stems *not* ribbed (rarely somewhat so in *Opuntia Stanlyi);* fruit *not* bearing scale-leaves, indehiscent.   1. *Opuntia,* page 29

1. Glochids none; stem leafless or with only minute bulges or scales representing leaves; floral tube often developed into a deep cup or a tube above the ovary and below the perianth parts; seeds *not* bony.

  2. Flowers and fruits produced on parts of the stem at least one and often several years old, *not* on the new growth of the current season, usually clearly below the stem-apex; seed longer than broad, therefore the hilum obviously basal or rarely oblique.

    3. Stem ribbed by coalescence of the tubercles; flower bud *either* developing within the spine-bearing areole *or* bursting through the epidermis just above it; ovary of the flower and fruit bearing scale-leaves or spines.

      4. Flower bud within at least the edge of a mature spine-bearing areole; stem length 15 to 100 times the diameter; plant never depressed or caespitose; the mature stem(s) 1 to 50 feet long, when shorter than 2 feet, less than 1 inch in diameter.   2. *Cereus,* page 107

      4. Flower bud bursting through the epidermis of the stem just above a mature spine-bearing areole; plant depressed to caespitose; stems solitary or several or in some species often numerous and forming a mound, 2 to 12 or 24 inches long, those of maximum length 2 to 4 inches in diameter.
                3. *Echinocereus,* page 120

    3. Stems *not* ribbed, with independent tubercles; flower bud in a special areole or area between one tubercle and the next one above it, *not* associated with the spiniferous areole on the tubercle or with any apparent extension of it; ovary of the flower and fruit bare.   4. *Mammillaria,* page 148

  2. Flowers and fruits produced on new growth of the current season, therefore near the apex of the stem or branch (except in *Coryphantha recurvata),* borne on the side of the tubercle toward the growing apex of the stem, either in a felted area adjacent to and merging with the new spine-bearing areole or in a similar remote area on the side or at the base of the tubercle but connected with the main areole by a felted groove (isthmus), the flowering portion and the groove (if any) persisting for many years as a circular to jagged scar either just below the area of spines or extending down the ventral side of the tubercle.

  3.' Flower-bearing portion of the areole adjacent to and merging into the spine-bearing portion; flower or fruit crowded against the edge of the spine-cluster; the scar remaining after fall of the fruit connected directly with and merging into the spine-bearing area, not separated by an isthmus, one-eighth to one-half or rarely (sometimes in *Sclerocactus)* extending the full length of the tubercle.

    4.' Stems of mature plants strongly ribbed; rib formed by coalescence of the bases of tubercles, the free upper portion of the tubercle from one-quarter as high as the rib beneath it to of equal height; ovary with numerous scale-leaves; fruit indehiscent, thick-walled.

      5. Mature fruit not splitting open; ovary *either* fleshy and scaly *or* dry and covered and obscured by long wool.

        6. Fruit fleshy for several months after reaching maturity, the areoles of the fruit *not* bearing long wool; sepaloid perianth parts *not* spinose or aristate; seed longer than broad, the hilum obviously basal or "sub-basal" or "diagonal."   5. *Ferocactus,* page 163

        6. Fruit (at least in the spiny species) becoming dry soon after maturity; the areoles bearing long woolly hairs which obscure the ovary; sepaloid perianth parts with spinose or aristate tips; seed broader than long, the hilum appearing "lateral."
                6. *Echinocactus,* page 172

      5. Mature fruit dry, releasing the seeds *either* by (1) separating crosswise at some level *above* the base to near the middle *or* by (2) opening along *two or three* vertical slits; ovary *either* (1) without scales *or* (2) with only a few scales, the areoles axillary to these bearing inconspicuous tufts of short hairs; seed broader than long, the hilum therefore appearing "lateral," 1/10 or 1/8 to 1/6 inch long; flower-bearing portion of the areole sometimes elongate and from one to several times as long as broad, the scar sometimes extending even the full length of the tubercle.
                7. *Sclerocactus,* page 175

4.′ Stems *not* ribbed, the tubercles separate; ovary and fruit with few or no scales, the surface never obscured by hairs from the areoles; seed broader than long, the hilum appearing "lateral."

5.′ Fruit dry and green or brown at maturity, bearing one to several scale-leaves, *either* (1) the length and diameter about equal *or* (2) the length up to 1½ times the diameter, the fruit at maturity *both* splitting open along the dorsal (back) side *and* with the top and the dried floral parts lifting off like a lid.
8. *Pediocactus,* page 179

5.′ Fruit fleshy but the wall thin and a brilliant red at maturity, without scale-leaves, the length several times the diameter, not splitting open, the floral cup deciduous or nearly so.
9. *Epithelantha,* page 189

3.′ Flower-bearing portion of the areole (except on juvenile stems) remote from the spine-bearing portion; flower or fruit apart from the spine-cluster, the scar left after the fall of the fruit depressed, at the end of a very narrow felted and usually line-like groove extending to the base of the tubercle (except on a juvenile stem or its persistent tubercles or ribs at the base of an older stem *).

4.″ Ovary bearing one to many scale-leaves, the fruit dry at maturity, opening *either* (1) lengthwise *or* (2) diagonally at the base and releasing the seeds.
10. *Neolloydia,* page 189

4.″ Ovary *not* bearing any scale-leaves, the fruit green or red, fleshy but very thin-walled at maturity, not splitting open.
11. *Coryphantha,* page 194

* The earliest flowers on the young stem in at least some species of *Coryphantha* and also in related genera not occurring in Arizona are produced high on the tubercle (which is separate and not a part of a rib). From then on, as in later seasons flowers appear on new crops of tubercles, they develop lower and lower on the upper side of each tubercle. The connecting grooves therefore become longer and longer. Finally the flower comes from the base of the tubercle, and the groove extends the full length.

1. OPUNTIA
Prickly Pear, Cholla

Stems composed of series of joints, the larger ones cylindroidal or flattened; ribs none, the tubercles separate. Leaves cylindroidal, acicular, or subulate, mostly ¼ to 1 or infrequently 2 inches long. Spines smooth, white, gray, yellow, brown, red, pink, or purplish, commonly 1 to 10 or 15 per areole but sometimes none; glochids (minute, sharp-pointed, barbed bristles — a feature of only this genus) produced in the areoles with the longer spines. Flowers and fruits produced on the joints grown in the preceding season, located near the apex of the joint, developed within the spine-bearing areole. Floral tube above its junction with the ovary very short, deciduous after flowering, bearing stamens just above the ovary. Fruit fleshy or dry at maturity, with or without spines or hairs from the areoles, with the floral tube deciduous, indehiscent. Seeds gray, tan, brown, white, or tinged with other colors, flat and bony, smooth or irregularly angled or indented, either longer than broad (length being hilum to opposite side) or discoid or nearly so; cotyledons leaflike.

DISTRIBUTION: Many species occurring in the Western Hemisphere, the number uncertain and doubtless grossly overestimated in all publications. Forty-seven native or introduced species occurring in the United States. Twenty-six in Arizona.

KEY TO THE SUBGENERA

1. Joints of the stem cylindroidal, circular in cross section; epidermis of the spine separating during the first year of growth into a thin, paperlike sheath, in low mat-forming species the sheath separating at only the tip; glochids *usually* small and inconsequential, except on mostly the underground stems and the fruits of the mat-forming species.

Subgenus 1. *Cylindropuntia*, page 30

1. Joints of the stem, excepting those formed in first-year seedlings, flattened; spine with *no* sheath; glochids usually large, well-developed, barbed, and effective.

Subgenus 2. *Opuntia*, page 70

Subgenus 1
CYLINDROPUNTIA Engelm.
Cholla

Joints of the stem cylindroidal or at least circular in cross section. Spine with the epidermis either completely or incompletely separating into a thin, paperlike sheath, sometimes in low mat-forming species, the sheath separating at only the tip; glochids usually small and inconsequential, except on the underground stems and the fruits of some mat-forming species.

DISTRIBUTION: Twenty-one species in the United States from Nevada and southern California to Colorado and southern Texas. Some of the same species ranging into Mexico; other species in Mexico, Central America, the West Indies, and South America.

KEY TO THE SECTIONS

1. Spine during the first year of development with the epidermis separating and forming a thin, paperlike sheath, slender, not papillate or striate; glochids nearly always all small and harmless; joints cylindroidal, usually uniformly spiny, of varying lengths, developing from any areole of an older joint; terminal bud continuing to grow and the joint to elongate for an undetermined period, the plant therefore shrubby, arborescent, or treelike.

Section 1. *Cylindropuntia*, page 31

1. Spine with the epidermis separating into a sheath at only the apex, the larger spines flattened and with either rough cross bands or papillae or longitudinal ridges-and-grooves; glochids (especially on underground stems and often on fruits) large and strongly barbed; joints enlarged upward like clubs, all about the same length, new ones developing only from the basal areoles of the older joint, the plant therefore forming a mat; terminal bud ceasing to grow when joint reaches a particular size.   Section 2. *Clavatae,* page 63

Section 1
CYLINDROPUNTIA
Chollas

Shrubs, bushes, arborescent plants, or small trees, but sometimes thicket-forming, mostly 1 to 15 feet high; branches developing from any areole; joints theoretically of unlimited size, with the terminal bud growing for more than a predetermined period and adding length to the joint, cylindroid, attached firmly or readily detached, uniformly spiny; sheaths of the spines thin, deciduous, separating through their entire length, not banded with papillae or ridged-and-grooved, the largest differing only in size from the rest; glochids almost always small and inconspicuous. Fruit fleshy or dry at maturity.

DISTRIBUTION: Seventeen species native in the western United States from California to Colorado and Texas; others from Mexico to South America and the West Indies.

KEY TO THE SERIES

1. Stem smooth or with simple tubercles.
    2. Fruit dry when the seeds reach maturity, spiny or at least sparsely so, the spines of the fruit in all but one very rare species strongly barbed.
        Series 1. *Echinocarpae,* page 31
    2. Fruit fleshy at maturity or drying after long persistence on the plant, spineless or rarely slightly spiny.
        3. Terminal joints or at least some of them not less than ⅝ inch in diameter.
            4. Larger terminal joints ⅝ to 1 or sometimes 1¼ inches in diameter at maturity, attached firmly, often elongate; spines of various colors, but often dark.
                Seires 2. *Imbricatae,* page 36
            4. Larger terminal joints 1¼ or 1½ to 2 inches in diameter at maturity, readily

detached and vegetatively reproductive; spines usually straw-color.
                Series 3. *Bigelovianae,* page 46
        3. Terminal joints not more than ½ inch in diameter, *not* markedly woody in the first year of growth; spines *not* strongly barbed.
            Series 4. *Leptocaules,* page 55
1. Stem nearly covered with flattened, plate-like, diamond-shaped to obovate tubercles; areole in an apical groove or notch of the tubercle; fruit dry at maturity, spiny, brown; joint becoming woody during the first year; spines quite strongly barbed.
        Series 5. *Ramosissimae,* page 61

Series 1
ECHINOCARPAE
Dry-fruited Chollas

Erect or rarely prostrate, usually shrubby, intricately branched chollas commonly 1½ to 6 feet high, some branches as long as the main trunk, if any; larger terminal joints ⅝ to 1¼ inches in diameter, with prominent raised tubercles, attached firmly, becoming woody in the first year. Fruit dry at maturity, obovoid to almost hemispheroid, depressed apically, deciduous immediately, the spines largely in an apical ring and (except in one species) strongly barbed.

DISTRIBUTION: Four species occurring in the United States from southern California to southwestern Utah and northwestern and southern Arizona. Mexico in northern Baja California and Sonora.

KEY TO THE SPECIES

1. Terminal joints ¼ to ⅜ inch in diameter; central spine 1, or sometimes also 1 or 2 shorter upper ones, the radial spines 6 to 8 and much smaller; fruit with several spines per areole, these weakly barbed; southeastern California and southwestern Arizona.
            1. *Opuntia Wigginsii,* page 32
1. Terminal joints or some of them at least ⅝ inch in diameter; central and radial spines indistinguishable; spines of the fruit strongly barbed.
    2. Tubercles once or twice as long as broad; longer terminal joints 4 to 6 inches long; trunk distinctly developed, often one-third to one-half the height of the plant.
            2. *Opuntia echinocarpae,* page 32
    2. Tubercles three to several times as long as broad; longer joints 6 to 18 inches long; trunk either none or rarely more than one-fifth the height of the plant.
            3. *Opuntia acanthocarpa,* page 34

1. Opuntia Wigginsii
L. Benson
sp. nov. (p. 19)

Shrub, 1 or 2 feet high, about 1 foot in diameter; trunk ¾ to 1½ inches in diameter; joints gradually expanded upward, 2 to 4 inches long, ⅜ inch in diameter; tubercles up to 3/16 inch long; spines moderately dense but not obscuring the joint, red or pink but with straw-colored sheaths, 6 to 8 per areole, those on the terminal part of the joint much larger, the central one in the areole far larger than the others, ¾ to 1¾ inch long, basally about 1/96 to 1/48 inch in diameter, the radials almost hairlike, up to ¼ inch long, basally perhaps 1/240 inch in diameter, not markedly barbed; flower (judging by a dried one) about 1 to 1½ inches (2.5-3.8 cm.) in diameter; fruit green, dry at maturity, with all the spines of each areole well developed, barbed, rather flexible, the longer ones ⅜ to ¾ inch long, the fruit ⅝ to ¾ inch long, about ½ to ¾ inch in diameter; seeds tan, 3/16 inch long.

DISTRIBUTION: Sandy soils of small washes and flats in the lower desert at less than 1,000 feet elevation. Colorado Desert. California in eastern San Diego County and northeastern Imperial County. Arizona in Yuma County from the Quartzite area to the Gila River as far east as the edge of Maricopa County. This species occurs in out-of-the-way and overlooked areas, because "Cacti do not grow there."

TYPE COLLECTION: South of Quartzite, Yuma County, Arizona.

2. Opuntia echinocarpa
Engelm. & Bigelow
Silver or golden cholla

Much-branched shrub; trunk often forming one-third to one-half the height of the plant; joints 2 to 6 or rarely 10 to 15 inches long, ¾ to 1½ inches in diameter; tubercles conspicuous, 1 to 2 times as long as broad, ¼ to ⅜ or rarely ¾ inch long vertically; spines straw-colored, silvery, or golden (the sheaths of similar color, conspicuous and persistent), dense on the branches, about 3 to 12 per areole, spreading in all directions, straight, ¾ to 1½ inches long, up to 1/32 inch broad, subulate, not barbed; flower usually 1¼ to 2½ inches in diameter; pet-

Fig. 1.1. Silver cholla, *Opuntia echinocarpa.*

aloid perianth parts greenish-yellow, the outer ones usually with a tinge or streak of red, few, obovate-oblanceolate, up to 1 inch long, about ⅜ to ½ inch broad, rounded and mucronulate, essentially entire; fruit green but turning to light tan or straw color, with dense spreading spines on the upper half, obovoid-turbinate or nearly hemispheric, ¾ to 1 or 1¼ inches long, ½ to ¾ inch in diameter, seeds light tan, about ¼ inch in diameter.

DISTRIBUTION: Sandy or gravelly soil of slopes, flats, and washes. 1,000 to 4,000 or 5,600 feet elevation. Mojavean Desert and (less commonly) Colorado Desert. Arizona in Mohave, northernmost Coconino, Yuma, southwestern Maricopa counties. California in and near the Mojave and Colorado deserts; southern Nevada; southwestern Utah. Mexico in northeastern Baja California and northwestern Sonora.

According to A. A. Nichol, plants of the silver and golden forms remain distinct after transplanting to a uniform environment.

SYNONYMY: Opuntia echinocarpa Engelm. & Bigelow. *Cylindropuntia echinocarpa* F. M. Knuth. Type locality: "near the mouth of Williams' River." *Opuntia deserta* Griffiths. Type locality: Searchlight, Nevada.

Fig. 1.2. Silver cholla, *Opuntia echinocarpa*. The fruits have barbed spines, and they become dry at maturity of the seeds. (Photograph by Robert H. Peebles.)

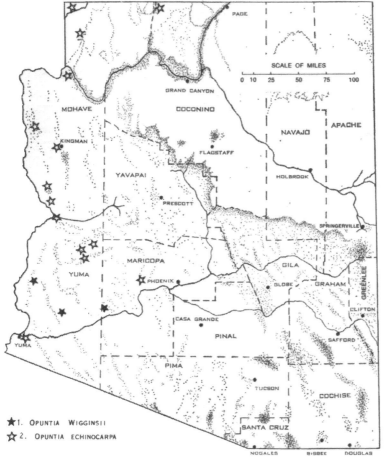

Fig. 1.3. The documented* distribution of *Opuntia Wigginsii* and of *Opuntia echinocarpa*.

* Except as noted, all locations on the maps indicate the source of a plant forming an herbarium specimen. The locations provide an approximate outline of distribution of the species or variety. Stippling indicates mountainous areas, that is, an irregular surface. The lower desert of Southern Arizona lies at a base level of 200 to 2,000 feet, northern Arizona (north of the Mogollon Rim) at a base level of about 5,000 to 7,000 feet.

3. *Opuntia acanthocarpa*
Engelm. & Bigelow
Buckhorn cholla

Shrub or an arborescent plant or a small tree; trunk short, less (usually much less) than one-fifth the height of the plant, up to 4 or 6 inches in diameter; joints 6 to 12 or 24 inches long, ⅝ or ¾ to 1¼ or rarely 1½ inches in diameter; tubercles conspicuous, sharply raised and laterally compressed, 3 to several times as long as broad, about ¾ to 1 or rarely 2 inches long vertically; spines tan to reddish-tan or straw-colored or whitish, turning to brown then to black in age (the sheaths conspicuous, straw-colored or rarely silvery, persistent about 1 year), 7 or 20 to 25 per areole, the longer ones ½ to 1 or 1½ inches long, 1/32 to 1/20 inch broad, subulate, not strongly barbed; flower 1½ to 2¼ inches in diameter; petaloid perianth parts variable in color, usually red, purplish, or yellow, narrowly obovate, 1 to 1½ inches long, up to ⅝ inch broad, rounded, mucronulate, with a few shallow sinuses; fruit turning to tan or brown, dry at maturity, tuberculate, except at the very base with numerous spreading spines, obovoid-turbinate, 1 to 1½ inches long, ⅝ to ¾ inches in diameter; seeds pale tan or whitish, irregularly angular, ¼ inch long.

3A. Var. **coloradensis**
L. Benson, **var. nov.** (p. 20)
Buckhorn cholla (Table 1)

DISTRIBUTION: Sandy or gravelly soils of benches, mountain slopes, and washes in the desert at 2,000 to 4,300 feet elevation. Mojavean and Sonoran deserts. Arizona from western Mohave County to northern Yuma County and eastward in the desert mountains to Yavapai and Gila counties. California in the eastern Mojave Desert and the eastern Colorado Desert; southern Nevada in western Clark County; southwestern Utah in Washington County.

The variety is named for its habitat, the basin of the lower Colorado River.

Fig. 1.4. Buckhorn cholla, *Opuntia acanthocarpa* var. *coloradensis* on tobosa grass flats near Aguila.

Fig. 1.5. Buckhorn cholla, *Opuntia acanthocarpa* var. *coloradensis*, the most common and widespread variety. The fruits have barbed spines, and they are dry at maturity of the seeds. (Photograph by Robert H. Peebles.)

3B. Var. **major**
   (Engelm. & Bigelow) L. Benson
   comb. nov. (p. 20)
   (Table 1)

DISTRIBUTION: Sandy soils of flats and washes in the desert at 1,000 to 3,000 feet elevation. Arizona Desert. Arizona from Yuma County to Gila, western Graham and western Pima counties. California north of Vidal Junction, Riverside County. Mexico in northern Sonora.

The Pima Indians steamed the flower buds in pits and used them for food.

SYNONYMY: Opuntia echinocarpa Engelm. & Bigelow var. *major* Engelm. & Bigelow. *Opuntia echinocarpa* Engelm var. *robustior* Coulter. Type locality: Sonora. *Opuntia acanthocarpa* Engelm & Bigelow var. *ramosa* Peebles. *Cylindropuntia acanthocarpa* (Engelm. & Bigelow) F. M. Knuth var. *ramosa* Backeberg. Type locality: Sacaton, Pinal County, Arizona.

3C. Var. *acanthocarpa*
   (Table 1)

DISTRIBUTION: Gravelly or sandy soils at or near the upper edge of the desert at about 4,000 feet elevation. Mojavean Desert. Arizona on the eastern edge of Mohave County and the western edge of Yavapai County from the Cottonwood Mountains to the McCloud Mountains. Utah near Virgin, Washington County.

SYNONYMY: Opuntia acanthocarpa Engelm. & Bigelow. *Cylindropuntia acanthocarpa* F. M. Knuth. Type locality: Cactus Pass, Yavapai County, Arizona.

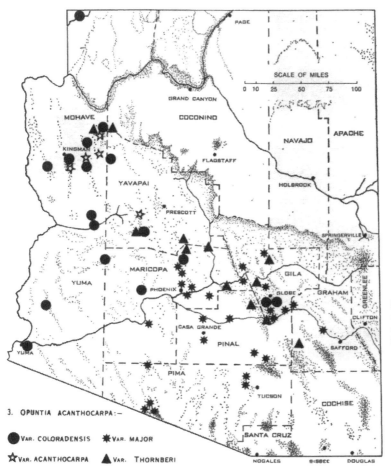

Fig. 1.6. The documented distribution of *Opuntia acanthocarpa*, according to its varieties.

3D. Var. *Thornberi*
    (Thornber & Bonker) L. Benson
    Thornber cholla (Table 1)

DISTRIBUTION: Rocky or gravelly soils of hillsides and ridges in the upper part of the desert at 2,500 to 3,500 feet elevation. Upper levels of the Mojavean and Arizona deserts. Arizona in the hills of the escarpment below the Mogollon Rim from Mohave County to Pinal and Gila counties.

SYNONYMY: Opuntia Thornberi Thornber & Bonker. *Opuntia acanthocarpa* Engelm. & Bigelow var. *Thornberi* L. Benson. *Cylindropuntia acanthocarpa* (Engelm. & Bigelow) F. M. Knuth var. *Thornberi* Backeberg. Type locality: Bumblebee, Yavapai County, Arizona.

Series 2
IMBRICATAE
Flowering Chollas

Erect shrubby or bushy plants, ½ to 8 feet high; main stem(s) branched and re-branched several times, some branches equal in length to the main trunk, if any; larger terminal joints mostly ⅝ to 1 or 1¼ inches in diameter, with prominent tubercles, attached firmly, becoming woody during the first year of growth. Fruit fleshy at maturity, green to red or purple or yellow, deciduous to persistent or in one species in southeastern Arizona proliferous (with one or two new fruits growing from the areoles of the old), spineless, tuberculate or smooth.

Fig. 1.8. Buckhorn cholla. *Opuntia acanthocarpa* var. *major.* The dry fruit has strongly barbed spines.

Fig. 1.7. Buckhorn cholla, *Opuntia acanthocarpa* var. *major,* the slender-jointed type common in central Arizona. The epithet *major* was applied when the plant was thought to be a large variety of *Opuntia echinocarpa.* Instead it is a small variety of *Opuntia acanthocarpa,* a much larger species. (Photograph by Robert H. Peebles.)

Fig. 1.9. Buckhorn cholla, *Opuntia acanthocarpa,* as described by Engelmann, not by recent authors. Note the broad tubercles. (From George Engelmann, "U.S. Senate Rept. Expl. & Surv. R. R. Route Pacific Ocean." *Botany 4:51. pl. 18. f. 1.* 1857; used by permission.)

DISTRIBUTION: Five species occurring in the United States from southern Nevada and Arizona to Colorado and western Texas. Other species occurring in Mexico.

KEY TO THE SPECIES

1. New flowers and fruits not formed in the areoles of persistent fruits of previous seasons; fruits *not* forming series with one growing from another, yellow, sometimes tinged with red or purple, strongly tubercled, each with a deep, cuplike cavity at the apical end; leaves either all very short or (the upper ones) slender and not more than 1/16 inch in diameter and ⅝ inch long.

2. Young spines with conspicuous white or silvery or light tan sheaths, these usually persistent after the first year of growth; larger terminal joints ½ to ¾ inch in diameter; leaves conical, 1/16 inch long; petals yellow; Utah and northern Arizona to southwestern Colorado and northwestern New Mexico.
4. *Opuntia Whipplei,* page 38

2. Young spines *not* with conspicuous white or silvery or light tan sheaths, the less conspi-

cuous sheaths usually deciduous after the first year of growth; larger terminal joints about 1 to 1¼ inches in diameter; leaves nearly cylindroidal, up to ½ to ¾ inch long; petals mostly reddish-purple, but sometimes yellow.

3. Tubercles ⅜ to ⅝ or ¾ inch long, 5 rows usually being visible from one side of the stem; spines ¼ to ⅜ inch long, moderately barbed; sheaths soon deciduous.
5. *Opuntia spinosior,* page 39

3. Tubercles about 1 inch long, 3 or 4 rows visible from one side of the stem; spines ¾ to 1¼ inches long, strongly barbed; sheaths persistent for about 1 year.
6. *Opuntia imbricata,* page 44

1. New flowers and fruits formed in the areoles of the persistent fruits of previous seasons; some fruits forming short series with one growing from the side of another, green, sometimes tinged with red, purple, or yellow, not markedly tubercled, with only a shallow cuplike cavity at the apical end; leaves elongate but not slender, up to ⅛ inch in diameter, up to ¾ inch long.   7. *Opuntia versicolor,* page 45

TABLE 1. CHARACTERS OF THE VARIETIES OF OPUNTIA ACANTHOCARPA

	A. Var. **coloradensis**	B. Var. **major**	C. Var. **acanthocarpa**	D. Var. **Thornberi**
Habit	Treelike or sometimes shrubby, the relatively few joints forming acute angles.	Shrub, sprawling or diffuse, the numerous joints forming acute and obtuse angles.	Shrub or arborescent plant, the relatively few joints forming acute angles.	Diffuse or low much-branched shrub, the relatively few joints forming acute angles.
Height	4 to 6 or 9 feet.	3 to 5 feet.	4 to 6 feet.	3 to 5 feet.
Joints Length Diameter	6 to 12 inches. 1 to 1¼ inches.	5 to 10 inches. ¾ to 1 inch.	8 to 20 inches. About 1¼ inches.	10 to 20 inches. About ¾ inch.
Tubercle Length Breadth	¾ to 1¼ inches. Narrow.	¾ to 1 inch. Narrow	1¼ to 1½ inches. Broad.	1¼ to 2 inches. Narrow.
Spine Number	10-12 per areole, dense on the fruits.	10-15 per areole, dense on the fruits.	12-20 per areole, dense on the fruits.	6-11 per areole, dense to sparse or none on the fruits.
Length	1 to 1½ inches.	1 inch.	1 inch.	½ to 1 inch.
Abundance	Moderate.	Moderate.	Sparse and not obscuring the stem.	Sparse and not obscuring the stem.
Geographical distribution	At higher levels. Arizona from Mohave to northern Yuma and Gila counties. California deserts to southwestern Utah.	At lower levels. Arizona in Yuma, Maricopa, Pinal and Pima counties.	Arizona along the border between Mohave and Yavapai counties. Utah in Washington County.	Arizona in the hills below the Mogollon Rim.

4. *Opuntia Whipplei*
   Engelm. & Bigel.
   Whipple cholla

Bushy or mat-forming or sometimes shrubby plants 1 to 2 or rarely 5 feet high; trunk none; stems erect, compactly arranged, with numerous short lateral branches; joints 1 or 3 to 6 inches long, ½ to ¾ inch in diameter; tubercles 2 to 3 times as long as broad, about ⅛ to ⅜ inch long, spines whitish-pink or pinkish-tan (the sheaths becoming loose and flattening out), conspicuous and white or silvery or rarely light tan or yellowish-pink, usually persisting one season, 7 to 14 per areole, mostly horizontal or deflexed, straight,

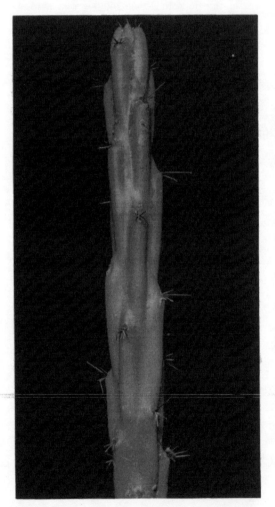

Fig. 1.10. Thornber cholla, *Opuntia acanthocarpa* var. *Thornberi*. Joint showing the remarkably long and slender tubercles.

up to ¾ to 1 or 2 inches long, basally up to about 1/32 inch in diameter, acicular, not strongly barbed; flower ¾ to 1¼ inches in diameter; petaloid perianth parts pale- or lemon-yellow or rarely greenish-yellow, narrowly obovate, ⅜ to ⅝ inches long, about ¼ to 5/16 inches broad, acute or obtuse and mucronate, somewhat crenate; fruit yellow, fleshy at maturity (except in parasitized fruits), strongly tuberculate, spineless, obovoid or subglobose, about ¾ to 1¼ inches long, ½ to ¾, or rarely ⅞ inch in diameter, with a deep, cuplike apical cavity (except in parasitized fruits), persistent through the winter, not developing new fruits from the sides of the old ones; seeds pale tan, about ⅛ inch long.

4A. Var. *Whipplei*
    (Table 2)

DISTRIBUTION: Usually in deep soils of valleys, plains, and gentle slopes of grasslands at 4,500 to 7,000 or occasionally 8,000 feet elevation. Great Plains Grassland and the borders of the Rocky Mountain Montane Forest, Juniper-Pinyon Woodland, and Sagebrush Desert. Arizona from Mohave County to Apache County and sparingly southward near the Mogollon Rim to Oracle in eastern Pinal County. Southern Utah; southwestern Colorado; northwestern and western New Mexico.

Over much of its range *Opuntia Whipplei* is a low mat-forming plant, but in other areas it is erect. In the rich valley bottoms of the Arizona Strip it attains a height of 6 or 7 feet. The fruit and young branches are a favorite food of antelope.

SYNONYMY: Opuntia Whipplei Engelm. & Bigelow. *Cylindropuntia Whipplei* F. M. Knuth. Type locality: Zuñi, New Mexico. *Opuntia Whipplei* Engelm. & Bigelow var. *enodis* Peebles. *Cylindropuntia Whipplei* (Engelm. & Bigelow) F. M. Knuth var. *enodis* Backeberg. Type locality: Hualpai Mountain, Mohave County, Arizona. Based upon insect-parasitized fruits. *Opuntia hualpaensis* Hester. *Cylindropuntia hualpaensis* Backeberg. Type locality: east of Peach Springs, Arizona.

4B. Var. **multigeniculata**
    (Clokey) L. Benson, **comb. nov.** (p. 20)
    (Table 2)

DISTRIBUTION: Rocky or sandy ridges at 4,700 feet elevation. Mojavean Desert. Arizona in Peach Springs Canyon, Mohave

Fig. 1.11. Whipple cholla, *Opuntia Whipplei,* growing at Prescott, Arizona Territory, September 27, 1904. (Photograph by David Griffiths.)

County. Nevada in Clark County east of Wilson's Ranch, Spring Mountains, west of Las Vegas; Washington County, Utah.

SYNONYMY: Opuntia multigeniculata Clokey. *Cylindropuntia multigeniculata* Backeberg. Type locality: Wilson's Ranch, Charleston [Spring] Mountains, Clark County, Nevada. *Opuntia abyssi* Hester. *Cylindropuntia abyssi* Backeberg. Type locality: Peach Springs Canyon, north of Peach Springs, Arizona.

5. *Opuntia spinosior*
   (Engelm.) Toumey
   Cane cholla

Small trees or shrubs 3 to 6 or up to 8 feet high; trunk short but sometimes up to 9 inches in diameter, the branches much longer; joints mostly 5 to 12 inches long, about ⅝ to ⅞ inches in diameter; tubercles numerous, not more than ⅜ inch apart, about 5 rows visible from one side of the stem, ⅜ to ⅝ or ¾ inch long; spines gray basally, pink above, with the sheaths dull tan and inconspicuous, about 10 to 20 per areole, spreading in all directions, straight, about ¼ to ⅜ inch long, up to 1/48 inch in diameter, acicular, barbed; flower about 1¾ to 2 inches in diameter; petaloid perianth parts purple or sometimes varying to shades of red, yellow, or rarely white, cuneate, ¾ to 1 inch long, ½ to ¾ inches broad, emarginate with a tooth in the notch, crenate-undulate; fruit bright lemon-yellow, fleshy at maturity, strongly tubercled,

TABLE 2. CHARACTERS OF THE VARIETIES OF OPUNTIA WHIPPLEI

	A. Var. **Whipplei**	B. Var. **multigeniculata**
Larger terminal joint length	3 to 6 inches.	1 to 2 inches.
Tubercles of terminal joints	2 to 3 or more times as long as broad, ⅛ to ⅜ inch long.	1½ times as long as broad, 3/16 or sometimes ¼ inch long.
Spines	4 to 7 per areole, sparse to moderately dense, ¾ to 1 or 1½ inches long.	10 to 14 per areole, densely crowded, ¾ to 1 inch long.
Sheaths	White or silvery or light tan or yellow.	Tan or yellowish-pink.
Color of petaloid perianth parts	Pale or lemon-yellow.	Light greenish-yellow.
Geographical distribution	Northern Arizona. Southern Utah to southwestern Colorado, and northwestern New Mexico.	Arizona in Peach Springs Canyon, Mohave County. Nevada at Wilson's Ranch, west of Las Vegas.

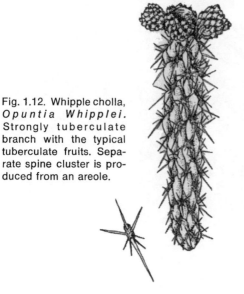

Fig. 1.12. Whipple cholla, *Opuntia Whipplei.* Strongly tuberculate branch with the typical tuberculate fruits. Separate spine cluster is produced from an areole.

spineless, obovoid, 1 to 1¾ inches long, ¾ to 1 inch in diameter, with the apical cavity deep and cuplike, persistent through the winter, new fruits not growing from the old; seeds light tan, 3/16 inch long.

DISTRIBUTION: Deep soils of valleys and plains or hillsides; mostly in grasslands at 2,000 to 6,500 feet elevation. Southwestern Oak Woodland and Chaparral and occasionally at the lower edge of the Rocky Mountain Montane Forest; typical of the Desert Grassland; upper parts of the Arizona Desert. Arizona from the Bradshaw Mountains and Agua Fria River Valley southward and eastward to Pima, Santa Cruz, and Cochise counties. New Mexico from Valencia County to Hidalgo and Luna counties and rare eastward to Socorro and Doña Ana counties. Mexico in Sonora and Chihuahua.

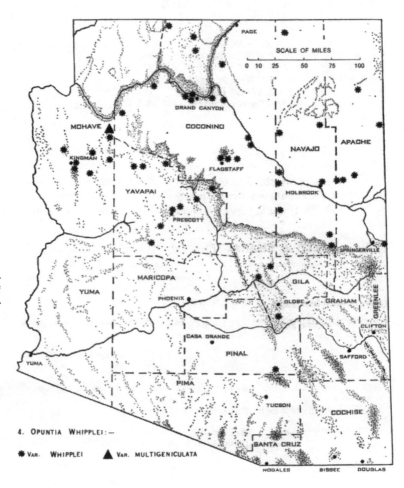

Fig. 1.13. The documented distribution of *Opuntia Whipplei,* according to its varieties.

Fig. 1.14. Cane cholla, *Opuntia spinosior*. (Photograph by David Griffiths, small range tract, Tucson, May, 1901. From David Griffiths and R. F. Hare, Prickly Pear and Other Cacti as Food for Stock II. New Mexico College of Agriculture and Mechanic Arts Bulletin No. 60, fig. 1, 1906. Used by permission.)

Fig. 1.15. Cane cholla, *Opuntia spinosior*. Branches and the strongly tuberculate fruits. (Photograph by Robert H. Peebles.)

The gray or purplish-gray of the spines makes this species conspicuous, as does production of whorls of short joints spreading at right angles to the main stems. The variation of flower color is considerable.

The fruits become yellow during the winter, and they fall readily about March, unless animals find them earlier.

Canelike branches are used for canes and other trinkets carved from the meshwork of woody tissue. Decay of the soft tissues of dead branches commonly leaves netted hollow cylinders of wood on the desert.

* * *

A plant collected at Tucson, and photographs of others along the Gila River in Pinal County near Sacaton, may represent hybrids of this species and *Opuntia fulgida*. The Sacaton plants are intermediate in joint diameter, length and color of spines, color of fruit, and ability to produce chains of fruits like those of *Opuntia fulgida*. The fruit is markedly tuberculate, and the apex is a deep cup as in *Opuntia spinosior*, but the tubercles are mammillate and the petals are few as in *Opuntia fulgida*. The drooping branches and the compact crown are intermediate but somewhat like those of *Opuntia fulgida* var. *mammillata*.

Plants growing in New Mexico intergrade with *Opuntia imbricata*. One reported from Graham County, Arizona, was intermediate between *Opuntia spinosior* and *Opuntia imbricata*.

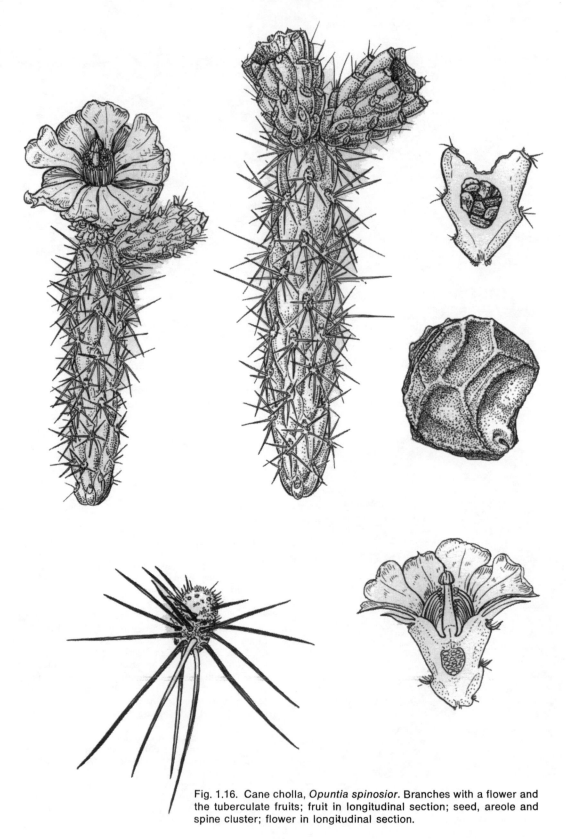

Fig. 1.16. Cane cholla, *Opuntia spinosior*. Branches with a flower and the tuberculate fruits; fruit in longitudinal section; seed, areole and spine cluster; flower in longitudinal section.

Fig. 1.17. Chollas in flower. *Above*, cane cholla, *Opuntia spinosior*. *Below*, staghorn cholla, *Opuntia versicolor*. (Photographs by William P. Martin.)

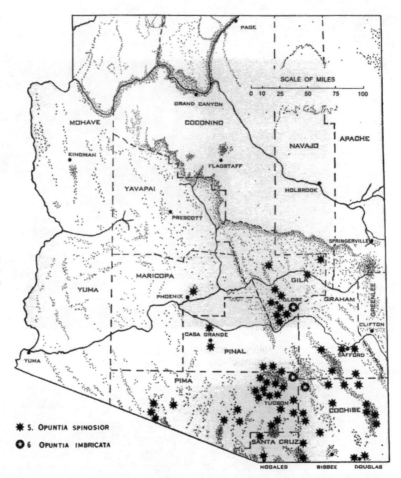

Fig. 1.18. The documented distribution of *Opuntia spinosior* and of *Opuntia imbricata.*

6. *Opuntia imbricata*
   Haw.
   Tree cholla

Small tree, arborescent plant, or thicket-forming shrub, usually 3 to 7 or sometimes 10 feet high; trunk short, branches much longer; joints 5 to 15 inches long, about ¾ to 1 inch in diameter; tubercles relatively few as compared with *Opuntia spinosior* but exceedingly prominent, very sharply raised, about 3 or 4 row visible from one side of the stem, ¾ to 1⅜ inches long; spines red or pink, with the sheaths dull tan and papery and persistent about one year, about 10 to 30 per areole, spreading in all directions, straight, ½ to 1⅛ inches long, about 1/48 inch broad, nearly acicular but somewhat flattened, strongly barbed (being less so in *Opuntia spinosior* and only slightly so in the other spe-

cies of the series); flower about 2 to 3 inches in diameter; petaloid perianth parts reddish-purple, narrowly obovate, 1 to 1½ inches long, ½ to ¾ inch broad, apically broadly rounded, entire to somewhat crenate-undulate; fruit yellow, fleshy at maturity, strongly tuberculate, spineless, obovoid, 1 to 1¾ inches long, ¾ to 1¼ inches in diameter, with the umbilicus deep and cuplike, persistent through the winter, young fruits not growing from the sides of the old; seed ⅛ inch in diameter.

DISTRIBUTION: Gravelly or sandy soils of hills and plains mostly in grassland at 4,000 to 6,000 feet elevation. Great Plains Grassland, Juniper-Pinyon Woodland, and higher parts of the Chihuahuan Desert. Arizona in Gila, Pima, and Cochise counties

Fig. 1.19. Tree cholla, *Opuntia imbricata*. (Photograph by David Griffiths.)

(see map, Fig. 1.18). Colorado from near Colorado Springs southward and rare west of the Rocky Mountains in Archuleta County; New Mexico; Kansas (Richfield); Oklahoma (Kenton); the western half of Texas. North-central and central Mexico.

SYNONYMY: Cereus imbricatus Haworth. *Opuntia imbricata* DC. *Cylindropuntia imbricata* F. M. Knuth. Type locality: unknown. *Cactus cylindricus* James. Type locality: near the Rocky Mountains. *Opuntia arborescens.* Engelm. Type locality: Santa Fe, New Mexico. *Opuntia vexans* Griffiths. Type locality: Webb County, Texas.

7. *Opuntia versicolor*
Engelm.
Staghorn cholla

Small tree or arborescent plant or sometimes a widely spreading low shrub, 3 to 8 or 15 feet high; trunk short, the branches much longer; joints often elongate, mostly 5 to 14 inches long, ⅝ to ¾ inches in diameter; tubercles fairly prominent, three to five times as long as broad, ⅝ to 1 inch long; spines reddish or basally gray or apically yellow, with the inconspicuous sheaths grayish or yellowish and deciduous within a few months, about 7 to 10 per areole, spreading in all directions, straight, ¼ to ⅝ inch long, about 1/64 inch in diameter, acicular, not strongly

barbed; flower 1¼ to 2¼ inches in diameter; petaloid perianth parts red, lavender to magenta or rose-purple, yellow, green, bronze, brown, or orange (hence the name of the species), ¾ to 1 inch long, ⅜ to ⅝ inch broad, truncate to rounded, mucronate, undulate; fruit green with usually a tinge of purple or of lavender or red, fleshy at maturity, not strongly tubercled, usually spineless, obovoid, ¾ or 1 to 1½ or 1¾ inches long, about ¾ inch in diameter, with the apical cup shallow, persistent for more than one year, some new fruits developed from the areoles of older ones, thus forming short chains of two or three; seeds light tan, ⅛ to ⅕ inch long.

Fig. 1.20. Tree cholla, *Opuntia imbricata. Left,* immature fruits still bearing a leaf just beneath each areole of the inferior floral cup, which envelops and which is adherent to the ovary wall. *Above,* a branch with the mature, tuberculate, yellow fruits, each with a deep apical cup (umbilicus). *Below,* the young branches with the leaves still present, one below each areole. (Photograph by David Griffiths.)

Fig. 1.21. Staghorn cholla, *Opuntia versicolor*. (Photo by H. L. Shantz.)

DISTRIBUTION: Deep sandy soils of canyons, washes, and valleys in the desert at 2,000 to 3,000 feet elevation. Arizona Desert. Arizona in Pinal and Pima counties from Baboquivari Valley to the Santa Cruz River Valley; abundant in the foothills about Tucson. Common in Mexico in adjacent northern Sonora.

This distinctly treelike plant, having an openly branched, rounded crown at maturity, is one of the most striking chollas in southern Arizona. The purple or reddish color of the dark green branches is conspicuous, especially in cold weather or drought periods.

The species is variable in a number of respects, especially in flower color. The many intermediate forms between it and *Opuntia spinosior* are particularly abundant in the eastern part of the range of the species.

SYNONYMY: Opuntia versicolor Engelm. *Opuntia arborescens* Engelm. var. *versicolor* E. Dams. *Cylindropuntia versicolor* F. M. Knuth. Type locality: probably Tucson.

Series 3
BIGELOVIANAE
Piercing Chollas

Small trees or arborescent plants, usually 3 to 12 or 15 feet high; trunk often 3 to 7 inches in diameter; branches longer or shorter than the trunk; joints 1¼ or commonly 1½ to 2 inches in diameter at maturity; readily detached and then rooting, this being the chief method of reproduction, not woody during the first year of growth. Fruit fleshy at maturity, green, smooth to strongly tuberculate, fertile or in some species usually sterile, *either* deciduous *or* persistent and with new flowers and fruits formed each season from the areoles of the old fruits, spineless or essentially so.

DISTRIBUTION: Four species occurring in the United States. Coastal Southern California to central and southern Arizona. Northwestern Mexico in Baja California, Sonora, and Sinaloa.

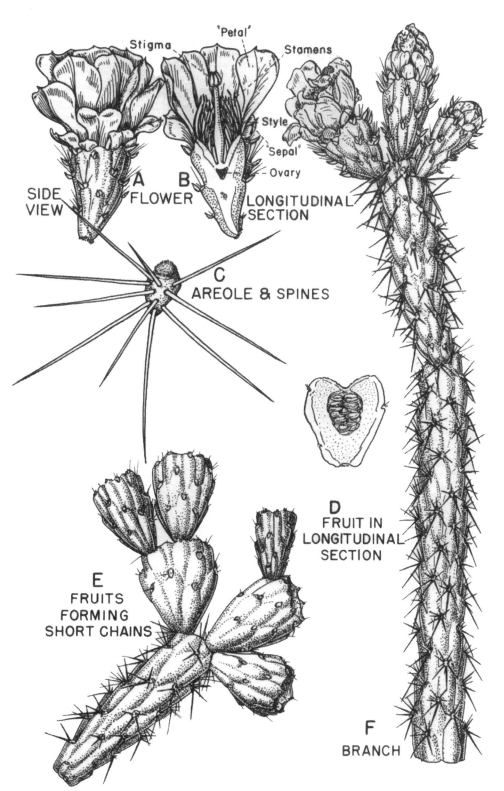

Stigma

"Petal"

Stamens

Style

"Sepal"

Ovary

A
SIDE VIEW

B
FLOWER

LONGITUDINAL SECTION

C
AREOLE & SPINES

D
FRUIT IN LONGITUDINAL SECTION

E
FRUITS FORMING SHORT CHAINS

F
BRANCH

Fig. 1.22. *A-C,* cane cholla, *Opuntia spinosior. E-F,* staghorn cholla, *Opuntia versicolor.*

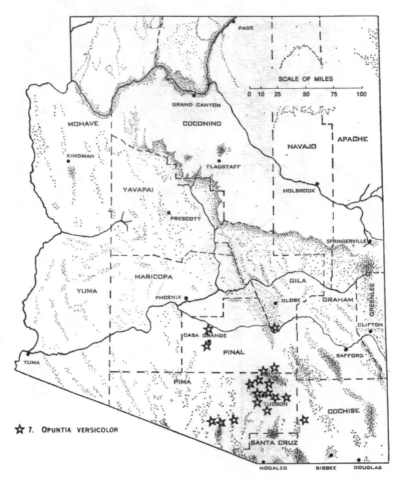

Fig. 1.23. The documented distribution of *Opuntia versicolor.*

☆ 7. OPUNTIA VERSICOLOR

Often reproduction is by rooting of the readily detached joints or, in some species, of the sterile fruits. This fixation of special combinations of genes through bypassing sexual reproduction is a means of maintaining a well-adopted pattern of hereditary characters successful for so long as the existing environment continues. Sexual reproduction would produce, instead, a variety of forms, many of them poorly adapted. However, with change of conditions, some plant types produced occasionally by sexual reproduction may survive and displace the prevailing types. The two varieties of *Opuntia fulgida* maintain themselves side by side through asexual reproduction. Their segregation doubtless is due to this factor.

KEY TO THE SPECIES

1. Fruits persistent and forming long branching chains up to 10-22 individuals in length, smooth or only slightly tubercled, with only a shallow apical cuplike depression (umbilicus), long-persistent; tubercles of the terminal joints distinctly longer than broad; stem branched below and usually several times rebranched above, with some of the main branches at least as long as the short main trunk; spines barbed but not strongly so.
   8. *Opuntia fulgida,* page 48

1. Fruits not forming chains, markedly tubercled, the apex of each with a deep cuplike depression, persisting only one season; tubercles of the terminal joints nearly as broad as long; stem forming a column, branching on only the upper portion, the branches rebranched only one or two times, the branches much shorter than the main trunk; spines effectively and strongly barbed.
   9. *Opuntia Bigelovii,* page 50

8. *Opuntia fulgida*
   Engelm.
   Jumping cholla

   Small trees or sometimes arborescent plants, 3 to 12 or 15 feet high; trunk 1 to 2 or 3 feet high, branched and rebranched several times, the branches being longer than the trunk; joints cylindroid or narrowly ellipsoid, 2 to 6 inches long, 1¼ to 2 inches in dia-

Fig. 1.24. Jumping cholla, *Opuntia fulgida*. Readily detached branches fall to the ground *(A)* and grow into new plants, often in profusion *(B)*. Photos: *A* by R. B. Streets; *B* by J. G. Brown.)

meter, readily detached, rooting, this being the principal method of reproduction, not woody during the first year of growth; tubercles mammillate, large, ½ to ⅞ inch long; spines either dense and conspicuous and tending to obscure the stem or more rarely (in var. *mammillata*) sparse, pink or reddish-brown, with the sheaths loose and of markedly greater diameter than the spines, 2 or 6 to 12 per areole, spreading in all directions, straight, ½ or ¾ to 1⅛ inches long, 1/48 to 1/42 inch broad, acicular to somewhat subulate, barbed but less so than in *Opuntia Bigelovii;* flower about ¾ inch in diameter; petaloid perianth parts pink or sometimes white with lavender streaks, only about 5 to 8, cuneate or cuneate-oblong, ¼ to ⅜ inch long, ¼ to ⅜ inch broad, apically rounded

and minutely denticulate; fruit green (usually sterile), fleshy at maturity, smooth, obovoid, without spines, 1 to 1½ inches long but continuing to grow after the first year, ¾ to 1 inch in diameter, the apical cup not deep, persisting up to at least twenty-five years, new ones growing from the old, forming long, branched chains; seeds light tan or yellowish, 3/16 inch long.

8A. Var. *fulgida*

Height 3 to 9 or 15 feet; younger joints not weak or drooping, with formation of considerable woody tissue after the first year; tubercles 3/16 to 1/4 inch high; spines about 6 to 12 per areole, ¾ to 1¼ inches long, strong, conspicuous, dense, and obscuring the joint.

This variety forms forests. The best remaining is on U.S. Highway 80 between Florence and Tucson. Much of this forest was partly destroyed by fire, but in part it has regenerated. A forest along Wilmot Road on the southeastern edge of Tucson was destroyed about 1940 to make an airfield. This is the most striking cholla in the United States. The loss of its elfin forests is to be regretted.

SYNONYMY: Opuntia fulgida Engelm. *Cylindropuntia fulgida* F. M. Knuth. Type locality: western Sonora.

8B. Var. *mammillata*
(Schott) Coulter

Height 2 to 4 feet; younger joints weak and drooping, almost lacking hard woody tissue, the formation of this delayed for some years; tubercles ¼ to ⅜ inch high, more prominent than in var. *fulgida;* spines 2 to 6 per areole, about ¼ to ½ or 1¼ inches long, slender, inconspicuous, sparse, and not obscuring the joint.

DISTRIBUTION: Sandy or gravelly soils of plains and hills at 1,000 to 2,500 feet elevation. Arizona Desert. Arizona in Pima County. Mexico in adjacent Sonora. The variety occurs with var. *fulgida* in many areas, and the few individuals are maintained as a separate population because of asexual reproduction. In some areas this is the dominant or exclusive type. The range of this variety is discontinuous but within the area of var. *fulgida.*

SYNONYMY: Opuntia mammillata Schott. *Opuntia fulgida* Engelm. var. *mammillata* Coulter. *Cylindropuntia fulgida* (Engelm.) F. M. Knuth var. *mammillata* Backeberg. Type locality: Baboquivari Mountains, Pima County, Arizona.

9. *Opuntia Bigelovii*
Engelm.
Teddy bear cholla

Miniature trees 3 to 5 or 9 feet high; branches much shorter than the trunk, rebranched only once or twice; trunk usually the height of the plant, becoming black as the spines turn to that color; mature joints green or bluish, narrowly ellipsoid, about 3 to 5 or 10 inches long, 1¼ or 1½ to 2½ inches in diameter, readily detached, mostly deciduous and rooting and reproducing the plant, leaving the main trunk like a post, not markedly woody during the first year of growth; tubercles 1 to 1.5 times as long as broad, six-sided, ¼ to ⅜ inch long and of

Fig. 1.25. Jumping cholla, *Opuntia fulgida,* showing the flowers and drooping branched chains of fruits. (Photographs by H. L. Shantz.)

DISTRIBUTION: Sandy soils of plains, mesas, washes, and hills in the desert at 1,000 to 3,000 or rarely 4,500 feet elevation. Arizona Desert. Southern Arizona from the Bill Williams River, northern Yuma County, and from Maricopa County to the lower parts of Gila County and southward to Pima County and northwestern Cochise County. Mexico as far southward as Sinaloa.

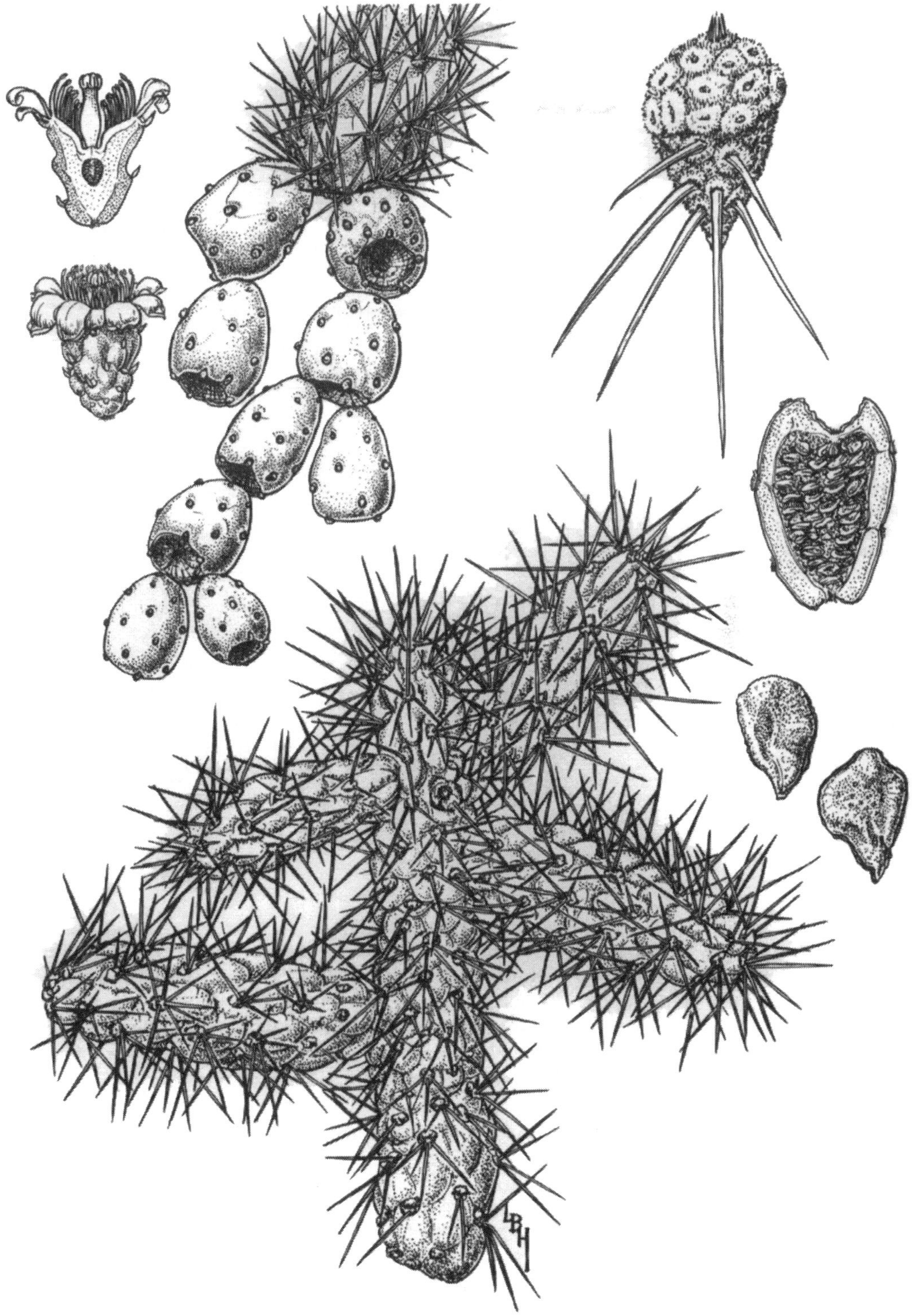

Fig. 1.26. Jumping cholla, *Opuntia fulgida. Upper left*, flower in longitudinal section and in external view. *Above*, branch tip with a drooping chain of fruits, these usually sterile and falling to the ground where they produce roots and grow into new plants. *Upper right*, enlargement of an areole and spines. *Right*, fruit in longitudinal section; enlarged seeds. *Below*, vegetative branches showing tubercles and spines.

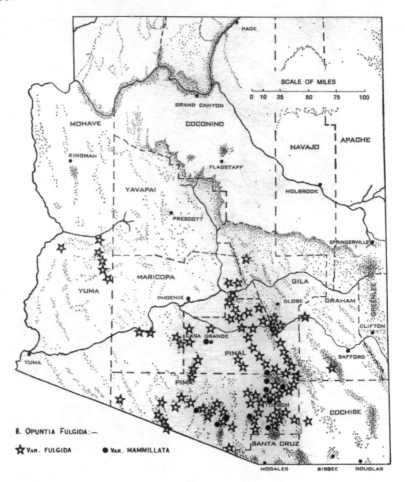

Fig. 1.27. The documented distribution of *Opuntia fulgida* according to its varieties.

equal breadth; spines conspicuous, dense, obscuring the joint, pinkish-tan or reddish-brown, with the sheaths usually straw-color and conspicuous and more or less persistent, about 6 to 10 per areole, spreading in all directions, straight, ⅝ to 1 inch long, about 1/24 inch broad, somewhat flattened, exceedingly strongly barbed, persistent for many years; flower 1 to 1½ inches in diameter; petaloid perianth parts pale green or yellow and streaked with lavender, narrowly cuneate-obovate, ½ to ¾ inch long, up to ⅜ inch broad, truncate, irregularly denticulate; fruit yellow or greenish-yellow (usually sterile), fleshy at maturity, strongly tuberculate, without appendages, ½ to ¾ inch long, ¼ to ⅜ inch broad, ½ to ¾ inch in diameter, the apical cup deep, persistent into or through the winter, not proliferous; seeds (when present) ⅛ inch long.

DISTRIBUTION: Rocky or gravelly areas of south-facing slopes of hills and mountains or on flats in the desert at 100 to 2,000 or 3,000 feet elevation. Colorado Desert and the lower portions of the Arizona Desert. Arizona from western Mohave and Maricopa counties southward to Yuma and Pima counties. California in the Colorado Desert. Mexico in northern Baja California and Sonora.

This never-to-be-forgotten cholla has a weird appearance and numerous, sharp, strongly barbed spines that can be removed from human skin and flesh only with great difficulty. Usually each spine must be clipped with scissors. *Opuntia fulgida* is of similar appearance, but it has much less strongly barbed spines.

SYNONYMY: Opuntia Bigelovii Engelm. *Cylindropuntia Bigelovii* F. M. Knuth. Type locality: Bill Williams River, Mohave County.

Fig. 1.28. *Opuntia fulgida* var. *mammillata.* Both varieties of the jumping cholla are maintained by asexual reproduction, both through joints that fall to the ground and through the usually sterile fruits that develop new plants similarly. The two types are maintained side by side with little or no interbreeding. Var. *mammillata* is more branched and compact and with a spreading, flat crown; its branches are weak and drooping and with fewer and shorter spines. The plant appears green instead of silvery or yellowish. (Photographs by H. L. Shantz.)

Fig. 1.29. Teddy bear cholla, *Opuntia Bigelovii*. *A,* plant with fallen (readily detached) joints, one of which has grown into a new plant. *B,* flower. (Photographs: *A,* by D. M. Crooks; *B,* by Robert H. Peebles.)

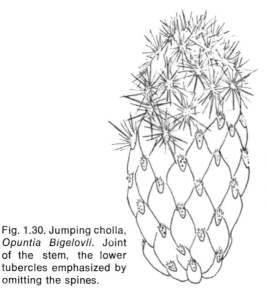

Fig. 1.30. Jumping cholla, *Opuntia Bigelovii*. Joint of the stem, the lower tubercles emphasized by omitting the spines.

Series 4
LEPTOCAULES
Pencil Chollas

Shrubby or bushy chollas or sometimes miniature trees, the branches slender and erect or diffuse, 1 to 2 feet or sometimes up to about 5 feet high; main branches several times rebranched, some branches as long as the main trunk (if any); joints not more than ¼ to ½ inch in diameter, nearly smooth or with simple, slightly-raised tubercles, attached firmly or readily deciduous, not woody during the first year of growth. Branches not densely spiny. Fruit fleshy at maturity, red or sometimes green with a purple tinge, deciduous, usually spiny.

DISTRIBUTION: Three species in the United States from Arizona to southcentral and southern Texas. These and others in Mexico, one being restricted to Baja California. One taxon occurring in the West Indies.

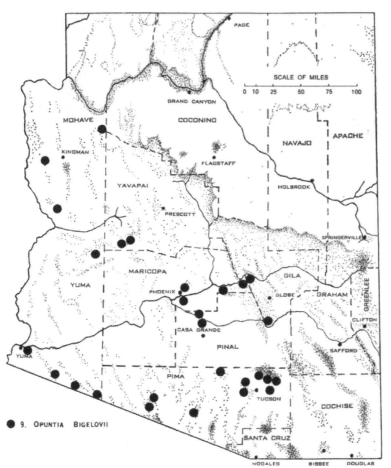

● 9. OPUNTIA BIGELOVII

Fig. 1.31. The documented distribution of *Opuntia Bigelovii*.

Fig. 1.32. Desert Christmas cactus, *Opuntia leptocaulis*. Long-spined form with the red fruits. (Photograph by David Griffiths.)

KEY TO THE SPECIES

1. Terminal joints 1/8 to 3/16 inch in diameter; fruits red, not with tubercles; plant without a trunk.   10. *Opuntia leptocaulis*, page 56

1. Terminal joints ¼ to ½ inch in diameter.

    2. Trunk 2 to 4 inches in diameter; fruit 1 to 1½ or 2 inches long, green but tinged with purple, lavender, or red, not markedly tubercled except in sterile or nearly sterile fruits; branches nearly smooth or with low tubercles 1 to 2 inches long.

                11. *Opuntia arbuscula*, page 58

    2. Trunk none; fruit ½ inch to ¾ inch long, red or mostly so, strongly tubercled; branches markedly tubercled, the strongly raised tubercles ½ to ⅝ or 1 inch long.

                12. *Opuntia Kleiniae*, page 59

10. *Opuntia leptocaulis*
DC.
Desert Christmas cactus

Bushes or erect small shrubs, often under protection of larger shrubs; trunk none; joints elongate, up to 12 to 16 inches long, the internal woody core nearly solid, the joints much shorter than the main branches, cylindroidal, 1 to 3 inches long, about 1/8 to 3/16 inch in diameter, the lateral joints short and sometimes at least at first spineless; tubercles almost lacking, the stem nearly smooth; spines gray, sometimes tinged with pink (the sheaths tan, conspicuous, persisting about one year, loose-fitting, of greater diameter than

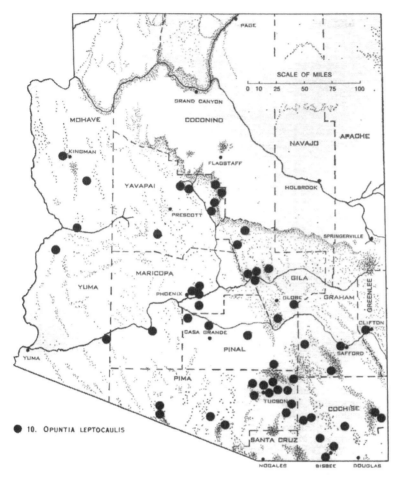

Fig. 1.33. The documented distribution of *Opuntia leptocaulis*.

the spines), 1 per areole, turned more or less downward, straight, the longer ones 1 to 2 inches long, circular in cross section, not markedly barbed; flower ⅜ to ⅝ inch in diameter; petaloid perianth parts green to yellow or bronze, obovate or cuneate-obovate, ¼ to ½ inch long, about ¼ inch broad, rounded, undulate; fruit bright red, fleshy at maturity, juicy, smooth, spineless, but the glochids often prominent, obovoid (or sterile fruits elongate), about ½ inch long, ⅜ to 7/16 inch in diameter, the apical cup usually shallow, persistent through the winter, sometimes with flowers and fruits developing from the areoles; seeds light tan, irregular, ⅛ to ⅙ inch long.

DISTRIBUTION: Mesas, flats, valleys, plains, and bottomland of washes in the deserts at 200 to 3,000 feet elevation. Mojavean, Arizona, and Chihuahuan deserts. Arizona from Hualpai Valley in Mohave County and Oak Creek Canyon in Coconino County southward to eastern Yuma County and to Greenlee, Pima, and Cochise counties. New Mexico from Valencia County to Quay County and southward; Oklahoma in the Arbuckle Mountains; Texas from El Paso to Oldham, Randall, Clay, Johnson, Burnet, Walker, and Victoria counties southward to the Rio Grande. Northern Mexico.

Among mesquites the species may become vine-like, growing upward through shrubbery for as much as 15 feet.

The bright red fruits are responsible for the localized English name, because they are attractive in winter when the desert is drab.

Fig. 1.34. Pencil cholla, *Opuntia arbuscula*. (Photograph by H. L. Shantz.)

SYNONYMY: Opuntia leptocaulis DC. *Cylindropuntia leptocaulis* F. M. Knuth. Type locality: " . . . in Mexico." *Opuntia frutescens* Engelm. var. *longispina* Engelm. *Opuntia leptocaulis* DC. var. *stipata* Coulter, *nom. nov. Opuntia leptocaulis* var. *longispina* Berger. *Cylindropuntia leptocaulis* (DC.) F. M. Knuth var. *longispina* F. M. Knuth. Type locality: "Laguna Colorado — East of the Pecos." Texas. *Opuntia vaginata* Engelm. *Opuntia leptocaulis* DC. var. *vaginata* S. Wats. *Cylindropuntia leptocaulis* (DC.) F. M. Knuth. Type locality: mountains near El Paso, Texas. *Opuntia fragilis* Nutt. var. *frutescens* Engelm. *Opuntia frutescens* Engelm. Type locality: Colorado River, Texas. The plants described as var. *frutescens* and var. *brevispina* have very short spines. However short-spined plants occur with long-spined almost throughout the range of the species. Detailed study of populations may reveal lines of segregation of varieties within *Opuntia leptocaulis,* but present information is meager. *Opuntia frutescens* Engelm. var. *brevispina* Engelm. *Cylindropuntia leptocaulis* (DC.) F. M. Knuth var. *brevispina* F. M. Knuth. Type locality: Texas.

11. *Opuntia arbuscula*
     Engelm.
     Pencil cholla

Miniature tree or a shrub, 2 to 4 or 9 feet high; trunk up to 1 foot long, 2 to 4 inches in diameter; joints 2 to 6 inches long, ¼ to ⅜ or ½ inch in diameter, the surface nearly smooth, the woody core an almost solid cylinder; tubercles low and inconspicuous, ¾ to 1¼ inches long; spines reddish- or purplish-tan (the sheaths light-brown, loose, of greater diameter than the spines, conspicuous), 1 or sometimes 2 to 4 per areole, the largest one turned downward, straight, up to ½ to 1½ inches long, basally flattened, narrowly elliptic in cross section, not markedly barbed; flower about ¾ to 1⅜ inches in diameter; petaloid perianth parts green, yellow, or terra cotta, obovate, ½ to 1 inch long, ⅜ to ⅝ inch broad, apically rounded, entire; fruit green tinged with purple or red, fleshy, not juicy, smooth, without spines, obovoid (or more elongate when sterile), ¾ to 1½ inches long, ½ to ⅞ inch in diameter, with the apical cup shallow, persistent through the winter, not giving rise to new flowers and fruits; seeds light tan, 3/16 inch long.

DISTRIBUTION: Sand and gravel in washes, flats, valleys, and plains in the desert at 1,000 to 3,000 feet elevation. Arizona Desert. Arizona from southern and eastern Yavapai County to Pima, western Santa Cruz, and western Cochise counties. Mexico in northern Sonora and Sinaloa.

Fig. 1.35. Pencil cholla, *Opuntia arbuscula.* Branches, the young one at the left with leaves, the two next to the left with fruits. (Photograph by David Griffiths.)

Fig. 1.36. Pencil cholla, *Opuntia arbuscula.* Branches with flower buds and flowers. (Photograph by David Griffiths.)

SYNONYMY: Opuntia arbuscula Engelm. *Cylindropuntia arbuscula* F. M. Knuth. Type locality: "On the lower Gila, near Maricopa, Arizona." *Opuntia vivipara* Rose. *Cylindropuntia vivipara* F. M. Knuth. Type locality: Tucson, Arizona. *Opuntia neoarbuscula* Griffiths. Type locality: Santa Rita Mountains, Arizona.

12. *Opuntia Kleiniae*
DC.
Klein cholla

Bushes or shrubs 1 to 7 feet high; stems branching divergently; joints with a purplish-red caste, 4 to 12 inches long, ¼ to ⅜ inch in diameter; tubercles prominent, ½ to ⅝ or 1 inch long; central core of the branch a nearly solid cylinder of wood, the pith forming only one-fourth to one-third the diameter of the stem; spines grayish-pink (the sheaths tan, loose, of much greater diameter than the spines, early deciduous), 1 to 4 per areole, mostly pointing downward, straight, up to ¾ to 1 or 1¼ inches long, slender, broadly elliptic in cross section, markedly barbed; flower 1¼ to 2 inches in diameter; petaloid perianth parts reddish to purple, cuneate to cuneate-obovate, ½ to ¾ inch long, ¼ to ⅜ inch broad, truncate or rounded, undulate

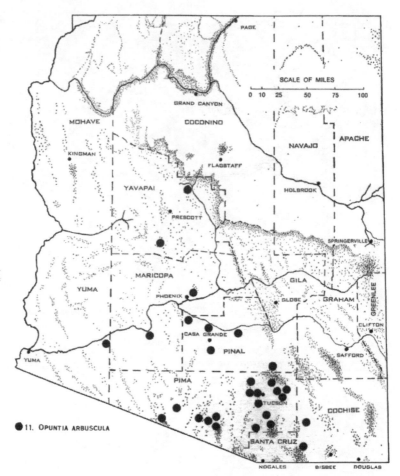

Fig. 1.37. The documented distribution of *Opuntia arbuscula*.

● 11. Opuntia arbuscula

to almost entire; fruit red (or sometimes green-and-red), fleshy but only slightly juicy, tubercled, spineless, obovoid, about ¾ inch long, ½ to ⅝ inch in diameter, with the apical cup shallow, persistent through the winter, not producing new flowers and fruits; seeds light tan, ⅛ inch long.

12A. Var. **tetracantha**
((Toumey) W. T. Marshall

Suberect or sprawling, 1 or 2 or sometimes 4 feet tall; each areole ¾ to 1 inch above the one directly below it; tubercles mostly about ⅝ inch long; spines ⅞ to 1¼ inch long, basally about 1/48 inch in diameter; petaloid perianth parts green but edged with red or brown.

DISTRIBUTION: Limestone flats and hills or along washes in the desert or grass-

land at mostly 2,000 to 3,000 or 4,400 feet elevation. Arizona sparingly from Yavapai County and southern Navajo County to Pima and Cochise counties. Mexico in Sonora and Sinaloa.

The distinction of varieties is largely the work of Pierre Fischer *(Taxonomic Relationships of Opuntia Kleiniae DeCandolle and Opuntia tetracantha Toumey.* Thesis. University of Arizona, 42. *f. 2, 9–10,* June 23, 1962). According to the cytological studies of Mr. Fischer, var. *tetracantha* is diploid, and var. *Kleiniae* of Texas is tetraploid.

SYNONYMY: Opuntia californica Engelm. Type locality: Arizona [not California] between the Coolidge Dam and Christmas. *Opuntia Thurberi* Engelm. Type locality: Bacuachi, Sonora, Mexico. *Opuntia tetracantha* Toumey. *Cylindropuntia tetracantha* F. M. Knuth. *Opuntia Kleiniae* DC. var. *tetracantha* W. T. Marshall. Type locality: east of Tucson, Arizona.

Series 5
RAMOSISSIMAE
Diamond Chollas

Bushy, matted, shrubby, or arborescent chollas, ½ to 2 or sometimes up to 5 feet high, the main branches rebranched profusely either at ground level or higher up, some branches being as long as the trunk (if any); joints up to ¼ inch in diameter, not readily detached, becoming woody the first year, covered with flattened, platelike, diamond-shaped to obovate tubercles, the areole in an apical notch or groove. Spines not obscuring the branches. Fruit dry at maturity, brown, subcylindroidal, deciduous, densely spiny, the spines up to ½ inch long.

DISTRIBUTION: One species occurring in deserts of the United States and north-

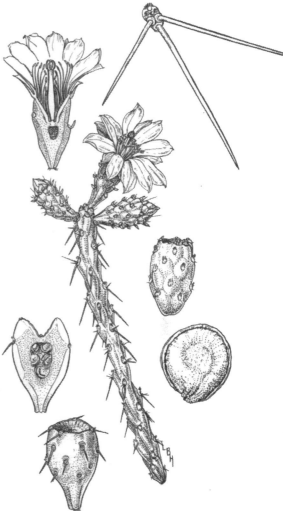

Fig. 1.38. A pencil cholla, *Opuntia Kleiniae* var. *tetracantha*. Branches and fruits. (Photograph by Robert H. Peebles.)

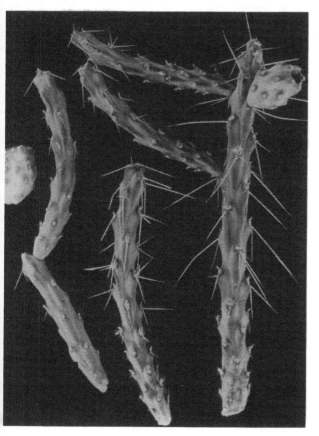

Fig. 1.39. A pencil cholla, *Opuntia Kleiniae* var. *tetracantha*. *Above,* a flower and an enlarged cluster of spines. *Center,* a flowering branch, showing the tubercles, areoles, and spines. *Below,* fruits, including a longitudinal section; an enlarged seed.

western Mexico. Southeastern California to Arizona. Mexico in northwestern Sonora.

13. *Opuntia ramosissima*
Engelm.
Diamond cholla

Bushy, matted, shrubby, or arborescent chollas ½ to 2 or 5 feet high; trunk rarely present; main branches rebranched profusely, some branches equalling the trunk (if any); joint grayish-green, slender, 2 to 4 inches long, ¼ inch in diameter; internal core of the

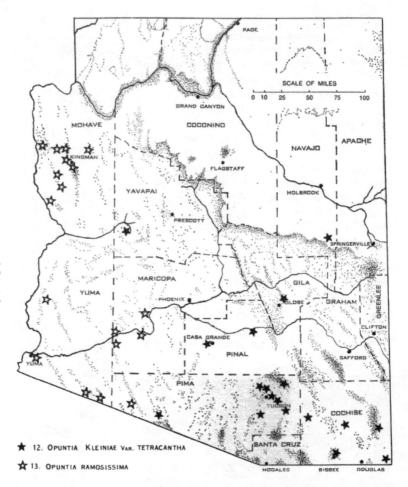

Fig. 1.40. The documented distribution of *Opuntia Kleiniae* var. *tetracantha* and of *Opuntia ramosissima*.

★ 12. OPUNTIA KLEINIAE VAR. TETRACANTHA

☆ 13. OPUNTIA RAMOSISSIMA

joint becoming woody the first year, nearly solid, not readily detached; tubercles flattened, platelike, diamond-shaped, the areole in an apical notch or groove, the tubercle 3/16 to 5/16 inch long; spines tan (the sheath light tan, apically reddish-tan, thin, membranous, conspicuous), in only the upper areoles, at first 1 to 3 per areole but only 1 developing, spreading or turned slightly downward, straight, the longer ones 1½ to 2¼ inches long, elliptic in cross section, with many minute strong barbs; flowers on short lateral branches, about ½ inch in diameter; petaloid perianth parts apricot to brown with some lavender or red, obovate, about ¼ inch long, up to ⅛ to 1/16 inch broad, sharply acute or acuminate, entire; fruit brown or tan, dry at maturity, densely spiny and burlike, (with spines up to ¾ inch long), ellipsoid, up to ¾ inch long, ½ inch in diameter, with

the apical cup deep; seeds creamy white or very light tannish-gray, ⅛ inch in diameter.

DISTRIBUTION: Sandy soils of washes and the desert floor at 100 to 2,000 or occasionally 3,000 feet elevation. Colorado Desert. Arizona from the lower areas of Mohave County to Yuma County and eastward to western Maricopa County and western Pima County. California in the southern portion of the Mojave Desert from the Mojave River to Death Valley and in the Colorado Desert; southernmost Nevada near the Colorado River. Mexico in northwestern Sonora.

The diamond-shaped cushionlike notched or grooved tubercles are a distinctive feature not duplicated in other chollas.

SYNONYMY: Opuntia ramosissima Engelm. *Opuntia tessellata* Engelm., *nom. nov. Cylindropuntia ramosissima* F. M. Knuth. Type locality: near the Colorado River in California.

Fig. 1.41. Diamond cholla, *Opuntia ramosissima*. Note the diamond-shaped tubercles. Each bears the areole in an apical groove. (Photograph by David Griffiths.)

Section 2
CLAVATAE
Engelm.*
Club Chollas

Mat-forming chollas 3 to 6 or 12 inches high; branching restricted to near ground-level, occurring only near the bases of the joints, these of limited size, the terminal bud ceasing to grow after the joint has reached its ultimate development, nearly always club-shaped, that is, enlarged upward from a narrow base, attached firmly, prominently spiny above but not below; spines with thin paper-like sheaths at only the extreme apices, at least the larger ones transversely banded with rough papillae or basally longitudinally ridged-and-grooved, the largest much exceeding the others, deflexed, flattened; glochids of

* *Clavatae* has precedence over *Corynopuntia*, which appeared in the first printing.

the underground stems and of the fruits much larger than those of the stem. Fruit fleshy at maturity, smooth, deciduous.

DISTRIBUTION: Four species in western North America from California to western Utah, southern New Mexico, and western Texas. Northwestern and north central Mexico.

KEY TO THE SPECIES
1. Perianth yellow or rarely reddish; larger spines rigid, papillate and rough.
    2. Spines *not* with ridges and grooves or only very faintly so (old disintegrating spines sometimes appearing so).
        14. *Opuntia Stanliy,* page 64
    2. Spines clearly ridged and grooved, about 1 inch long, basally proportionately very broad, subulate, daggerlike; northeastern Arizona and east of the Continental Divide in northern and central New Mexico.
        15. *Opuntia clavata,* page 67
1. Perianth reddish-purple; spines flexible, *not* papillate or rough, the larger ones usually with longitudinal basal ridges and grooves; Nevada to southwestern Utah and to northwestern Arizona.   16. *Opuntia pulchella,* page 68

14. *Opuntia Stanyli*
    Engelm.
    Devil cholla

Mat-forming cholla 6 inches or rarely 1 foot high, the mats up to several yards in diameter; joints cylindroid to narrowly obovoid, basally gradually or abruptly narrowed, 3 to 6 or sometimes 2 or 8 inches long, ⅝ to 1½ inches in diameter, enlarged upward, clublike, attached firmly; tubercles large and conspicuous, mammillate, ½ to 1¼ inches long; spines mostly on the upper portion of the joint, tan or straw-color to brown or red, the spines with rough papillae in crosswise ranks, not longitudinally ribbed or grooved, 10 to 18 per areole, the longest ones turned downward, straight, up to 2 inches long, basally 1/24 to 3/16 inch broad, linear-elliptic in cross section, not barbed; flower 1 to 2 inches in diameter; petaloid perianth parts yellow or sometimes reddish in var. *Parishii,* narrowly obovate, about ⅝ to 1 inch long, about ⅜ to ½ inch broad, attenuate; fruit yellow, fleshy at maturity, smooth, usually densely spiny, with large glochids, slender but enlarged upward, 1½ to 3¼ inches long, about ½ to ¾ inches in diameter, with a deep

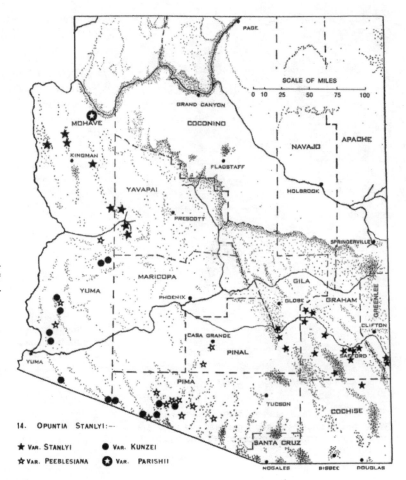

Fig. 1.42. The documented distribution of *Opuntia Stanlyi*, according to its varieties.

14.   OPUNTIA STANLYI:--

★ VAR. STANLYI   ● VAR. KUNZEI
☆ VAR. PEEBLESIANA   ✪ VAR. PARISHII

apical cup but this obscured by the persistent perianth, deciduous; seed light gray, tan, or yellow, ⅛ to ¼ inch long.

14A. Var. Stanyli
       Engelm.
       Devil cholla (Table 3)

DISTRIBUTION: Sandy or gravelly soils of plains, mesas, and washes in the desert at mostly 2,500 to 4,000 feet elevation. Upper part of the Arizona Desert and lower edge of the Desert Grassland. Southeastern Arizona from southeastern Pinal and southern Gila counties to southern Greenlee and northernmost Cochise counties. Southwestern New Mexico mostly near the Gila River in western Grant County and northern Hidalgo County.

SYNONYMY: Opuntia Stanlyi Engelm. *Corynopuntia Stanlyi* F. M. Knuth. Type locality: Upper Gila River Valley in Hidalgo County, New Mexico, near Virden.

14B. Var. **Peeblesiana**
       L. Benson, **var. nov.** (p. 20)
       Peebles cholla (Table 3)

DISTRIBUTION: Fine soils of valleys in the deserts at 1,000 to 2,000 feet elevation. Arizona Desert. Southern Arizona in Yuma (rare), southwestern Pinal, and western Pima counties from near the Bill Williams River to the Papago Indian Reservation south of Casa Grande and southward to Quitovaquita and the west side of Baboquivari Valley. Mexico in adjacent Sonora.

14C. Var. *Kunzei*
     (Rose) L. Benson
     Kunze cholla (Table 3)

DISTRIBUTION: Sandy or clay soils of valleys in the desert at 300 to 1,500 feet elevation. Colorado Desert. Western Arizona in Mohave, Yuma, and western Pima counties from Detrital Valley to the International Boundary and eastward to the Harquahala Mountains and the Organ Pipe Cactus National Monument. Mexico in Baja California between Mexicali and San Felipe and in Sonora as far south as St. George Bay.

Distinction as a species (under the name *Opuntia Wrightiana*) was based upon the striking extreme forms. Intergradation with var. *Peeblesiana* (cf. discussion there) is complete in a number of areas though not in others.

SYNONYMY: Opuntia Kunzei Rose. *Opuntia Stanlyi* Engelm. var. *Kunzei* L. Benson. *Corynopuntia Stanlyi* (Engelm.) F. M. Knuth var. *Kunzei* Backeberg. Name long wrongly applied. Type locality: Gunsight Mining Area, southeast of Ajo, Arizona. *Grusonia Wrightiana* Baxter. *Opuntia Wrightiana* Peebles. *Opuntia Stanlyi* Engelm. var. *Wrightiana* L. Benson. *Corynopuntia Stanlyi* (Engelm.) F. M. Knuth var. *Wrightiana* Backeberg. Type locality: Petrified Forest near the Colorado River, north of Yuma, Arizona.

14D. Var. *Parishii*
     (Orcutt) L. Benson
     Parish cholla (Table 3)

DISTRIBUTION: Sandy soils of valleys, plains, and mesas in the desert at 3,000 to 4,000 feet elevation. Mojavean Desert. Arizona in Mohave County (rare). Californian deserts in eastern San Bernardino and northern Riverside counties; southern Nevada in Clark County.

TABLE 3. CHARACTERS OF THE VARIETIES OF OPUNTIA STANLYI

	A. Var. **Stanlyi**	B. Var. **Peeblesiana**	C. Var. **Kunzei**	D. Var. **Parishii**
Joint series above ground level	Usually 1 or 2.	Usually 1.	Several.	Usually 1.
Joint length (largest joints)	3 to 6 inches or more.	3 to 6 inches.	4 to 6 or 8 inches.	2 to 3 inches.
Joint diameter	1 to 1½ inches.	⅝ to 1 inch.	1 to 1½ inches.	¾ to 1¼ inches.
Joint shape	Cylindroidal, abruptly narrowed basally.	Cylindroidal, gradually narrowed basally.	Cylindroidal, usually abruptly narrowed basally.	Obovoid or narrowly so, gradually narrowed basally.
Tubercles	Separate, 1 to 1¼ inches long.	Separate, ⅝ to 1 inch long.	In older plants basally coalescent into ribs, ¾ to 1¼ inches long.	Separate, ½ to 1 inch long.
Spines per areole	18-21, one much broader than the others.	23-26, *no* one much broader than the others.	26-33, *no* one much broader than the others.	16-20, one much broader than the others.
Seed length	3/16 to ¼ inch.	3/16 inch.	3/16 inch.	⅛ to 3/16 inch.
Flower color	Yellow.	Yellow.	Yellow.	Red or yellow.
Fruit	Spiny.	Spiny.	Spiny.	Usually not spiny, or weakly so.
Geographical distribution	Arizona from southeastern Pinal and Gila to Greenlee and northernmost Cochise counties. Southwestern New Mexico in the Gila River area.	Arizona in Yuma (rare), southwestern Pinal, and western Pima counties. Mexico in adjacent Sonora.	Arizona in Mohave, Yuma, and western Pima counties. Mexico in northern Sonora.	Arizona (rare) in Mohave County. California deserts, southern Nevada.

Fig. 1.43. Peebles cholla, *Opuntia Stanlyi* var.
Peeblesiana. (Photograph by Robert H. Peebles.)

B
AREOLE

A
BRANCH

C
SPINE

Fig. 1.44. Kunze cholla, *Opuntia Stanlyi* var. *Kunzei.*

SYNONYMY: Opuntia Parishii Orcutt. *Opuntia Stanlyi* Engelm. var. *Parishii* L. Benson. *Corynopuntia Parishii* F. M. Knuth. *Corynopuntia Stanlyi* (Engelm.) F. M. Knuth var. *Parishii* Backeberg. Type locality: Piñon or Pinyon Wells, southern edge of the Mojave Desert, California.

15. *Opuntia clavata*
Engelm.
Club cholla

Mat-forming cholla not more than 3 or 4 inches high, mostly 1 or 2 yards in diameter; larger terminal joints clavate, strongly expanded upward from narrow bases, attached firmly, mostly 1½ to 2 inches long, ¾ to mostly 1 inch in diameter, becoming woody during the first year; tubercles conspicuous and relatively large, mammillate, ½ inch long; spines mostly toward the upper part of the joint, ashy-gray, strongly papillate-roughened, clearly longitudinally ridged-and-grooved, 10 to 20 per areole, mostly deflexed, straight, about 1 inch long, ½ inch broad, strongly flattened, the largest one dagger-like, not barbed; flower about 1¼ to 2 inches in diameter; petaloid perianth parts yellow, cuneate or nearly obdeltoid, ¾ inch long, ½ inch or more broad, truncate or rounded; entire; fruit yellow, fleshy, smooth, covered with clusters of yellow or tan glochids, slender, enlarged upward, 1½ inches long, ¾ to ⅞ inch in diameter, with the apical cup very deep, deciduous; seeds yellow; a little less than ¼ inch long.

DISTRIBUTION: Sandy soils of valleys and plains in grassland at 6,000 to 8,000 feet elevation. Great Plains Grassland. Northern Arizona from the eastern edge of Coconino County to Apache County. The Arizona distribution is that mapped by A. A. Nichol from field studies for the first edition of this book (1940). It has not been substantiated by specimens, but the source is reliable. East of the Continental Divide in northern and central New Mexico from eastern Rio Arriba County to eastern Valencia, eastern Sierra, San Miguel, and northern Otero counties.

The species is remarkably distinctive. It is *not* closely similar to *Opuntia Stanlyi* var. *Parishii* as has been suggested.

SYNONYMY: Opuntia clavata Engelm. *Cactus clavatus* Lemaire. Type locality: Santa Fe, New Mexico.

Fig. 1.45. Club cholla, *Opuntia clavata*. (Photograph by David Griffiths.)

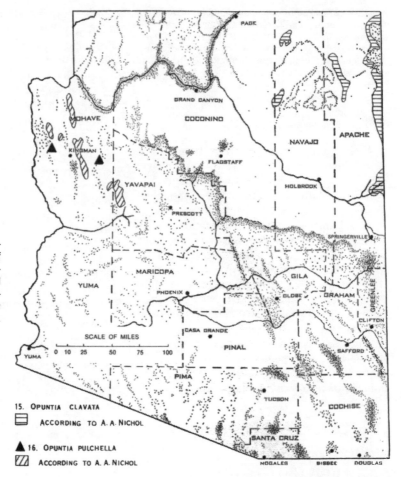

Fig. 1.46. The documented distribution of *Opuntia clavata* (as outlined by A. A. Nichol) and the documented distribution of *Opuntia pulchella* (also as outlined by A. A. Nichol.)

15. OPUNTIA CLAVATA
    ACCORDING TO A. A. NICHOL

16. OPUNTIA PULCHELLA
    ACCORDING TO A. A. NICHOL

16. *Opuntia pulchella*
    Engelm.
    Sand cholla

Inconspicuous clump-forming cholla, the clumps usually only a few inches in diameter; stems arising from a glochid-covered tuber about 2 or 3 inches in diameter, the areoles bearing the glochids which are in time deciduous; joints exceedingly variable, but especially so if the plants have been browsed by animals or if they are in poor health, expanded upward or narrowly ellipsoid or cylindroid, 1 to 1 1/2 or 2 1/2 or sometimes 4 inches long, 3/16 to 1/2 or 1 inch in diameter; tubercles inconspicuous or sometimes projecting and mammillate, 1/4 to 3/8 inch long; longer spines mostly toward the top of the joint, white to gray, brown, or pink, with basal ridges and grooves, not papillate, 8 to

15 per areole, mostly deflexed, straight but flexible, the longest one light-colored, up to 2½ inches long, up to about 1/16 inch broad, flattened, not barbed; in juvenile plants the spines all alike, white with tan bases, 1/16 to ⅛ inch long, very slender, not ridged or grooved; glochids, those below ground-level ⅜ to 1 inch long, those above ground yellow and usually much smaller and inconspicuous; flower ¾ or 1¼ to 1½ inches in diameter; petaloid perianth parts purple to rose, cuneate or cuneate-obovate, ⅝ to 1 inch long, ⅜ to ⅝ inch broad, truncate or retuse, mucronate, marginally undulate; fruit reddish, fleshy, smooth, the areoles prominent, each bearing 20 or more purplish or brownish bristlelike soft spines ½ to ⅝ inch long, the fruit about ¾ to 1⅛ inches long, ⅜ to ½ inch in diameter, with the apical cup very deep, sur-

Fig. 1.47. Sand cholla, *Opuntia pulchella*. The tuberous part of the stem below ground is covered at first by areoles with very large glochids, but eventually these areoles and their spines fall away as shown at the right. The branches protruding above the ground are slender, and bear elongate flattened spines. (Photograph by Robert C. Frampton Studios, Claremont.)

rounded by a flared structure derived from the floral cup, deciduous; seeds bone-white, 3/16 to 1/4 inch long.

DISTRIBUTION: Sand of dunes, dry lake borders, river bottoms, washes, valleys, and plains in the desert at about 4,000 to 5,000 feet elevation. Sagebrush Desert. Northwestern Arizona in Mohave County and, according to A. A. Nichol, the western edge of Yavapai County. Nevada from eastcentral Washoe County, Lyon County, and Esmeralda County to Lander and southern White Pine counties; western Utah.

The variety includes several minor forms or probably only abnormal types. Some of these forms have received names as species under the proposed genus *"Micropuntia"* based upon (1) presence of tubers without glochids (these, however, being deciduous in age with the complete areoles), (2) special spine types, these being highly variable in all populations and there being frequently abnormal types produced after injury, disease, or desiccation, and (3) small, often juvenile, joints, which are as variable as are the spine types. Cf. L. Benson, "The Opuntia pulchella Complex." Cactus & Succulent Journal 29:19–31. *3 photographs.* 1957.

The plants flower and fruit during juvenile stages, and this has given rise to confusion. Specimens collected by Gordon W. Gullion (Smoky Valley, Lander County, Nevada, in 1958, *Pom*) show transitions from juvenile to adult plants.

SYNONYMY: Opuntia pulchella Engelm. *Corynopuntia pulchella* F. M. Knuth. Type locality: Walker River, Nevada. *Micropuntia brachyrhopalica* Daston. *Opuntia brachyrhopalica* Rowley. Type locality: U.S. Desert Experimental Range west of Milford, Utah. *Micropuntia Barkleyana* Daston. Type locality: same as above. *Micropuntia spectatissima Daston.* Type locality: same as above. *Micropuntia tuberculosirhopalica* Wiegand & Backeberg. *Opuntia tuberculosirhopalica* Rowley. Type locality: "USA (Utah, Arizona)." *Micropuntia pygmaea* Wiegand & Backeberg. *Opuntia pygmaea* Rowley. Type locality: "USA (Idaho australis, Nevada)." *Micropuntia gracilicylindrica* Wiegand & Backeberg. *Opuntia gracilicylindrica* Rowley. Type locality: "USA (Nevada)." *Micropuntia Wiegandii* Backeberg. *Micropuntia gigantea* Wiegand ex Backeberg, *pro syn. Opuntia Wiegandii* Rowley. Type locality: "USA (Nevada, California)." Only the original *Opuntia pulchella* is documented by a specimen or, except for Daston's combinations, even attributed to a precise locality.

Subgenus 2
OPUNTIA
Prickly Pears

Joints strongly flattened. Spine *not* with the epidermis or any portion of it separating into a loose sheath; glochids well developed and markedly barbed.

DISTRIBUTION: Twenty-seven species native or introduced in the area from British Columbia to Ontario and Massachusetts and southward through the United States, especially in the Western and Plains states; a few native in the Southeast and especially in Florida; four or five species introduced; many native in Mexico, Central America, the West Indies, and South America.

Section 3
OPUNTIA
Prickly Pears

Trunk (when present) and the main branches and minor branches *all* composed of flattened joints, these growing to a predetermined size and the terminal bud ceasing further growth.

DISTRIBUTION: Twenty-five native or introduced species occurring in the United States. Eleven in Arizona.

KEY TO THE SERIES

1. Fruit becoming tan and dry as the seeds reach maturity in winter or spring; seeds usually ⅛ to 5/16 inch in diameter, rough and irregular. (Note: Sterile fruits and those parasitized by insect larvae may remain fleshy and greenish or slightly purplish.)

2. Fruit with at least an apical rim of divaricately spreading, strongly barbed spines, often spiny all over; joints glabrous, nearly always spiny.   Series 1. *Polyacanthae*, page 70
2. Fruit spineless or essentially so; joints pubescent in Arizona plants, spineless.
            Series 2. *Basilares*, page 81
1. Fruit fleshy and juicy at maturity, usually red or reddish-purple, nearly always spineless but with glochids; seeds usually small, ½ to 3/16 or sometimes ¼ inch in diameter, smooth and regular in outline.
            Series 3. *Opuntiae*, page 85

Series 1
POLYACANTHAE
Dry-fruited Prickly Pears

Low plants tending to creep or form mats, the clumps up to several yards in diameter, usually less than 6 or 8 inches high. Joints always with a bluish bloom (powdered wax), not hairy, nearly always spiny. Fruit becoming more or less tan or brown as the seeds mature, with at least an apical rim of widely spreading, strongly barbed spines, often spiny all over. Seeds large, usually ⅛ to 5/16 inch in diameter, the outer parts irregular and rough.

DISTRIBUTION: Five species in the western half of North America.

Sterile fruits and those parasitized by insect larvae may remain fleshy and greenish or slightly purplish.

KEY TO THE SPECIES

1. Spines *all* circular to elliptic in cross section, none markedly even basally flattened. (Note: Hybridization sometimes obliterates these distinctions in northern Arizona).
    2. Joints 2 to 4 or 5 inches long, 1½ to 4 inches broad, broadly obovate to orbiculate, up to about ½ inch thick, less than one-quarter as thick as broad; spines of the stems not strongly barbed; joints not readily detached.
            17. *Opuntia polyacantha*, page 70
    2. Joints 1 to 1½ inches long, 1 inch or less broad, flattened-obovoid or -ovoid, ½ to ¾ inch thick, at least one-half as thick as broad; spines strongly barbed; joints readily detached from the plant and clinging by the barbed spines.
            18. *Opuntia fragilis*, page 72
1. Spines *or some of them* elliptic or narrowly elliptic in cross section, being at least basally flattened.
    2.' Spines slender, the larger ones basally 1/48 or sometimes up to 1/72 inch broad, relatively flexible, sometimes slightly twisted; joints mostly 4 to 6 inches long and 2½ to 3½ inches broad, elliptic-oblong to obovate-oblong.   19. *Opuntia erinacea*, page 74

2.′ Spines stout, the larger ones basally 1/24 to 1/16 inch broad, only slightly flexible, the longer ones markedly twisted; joints 5 or mostly 8 to 10 inches long and 3 to 5 inches broad, obovate or narrowly so.
20. *Opuntia Nicholii,* page 80

17. *Opuntia polyacantha*
Haw.
Plains prickly pear

Prickly pears forming clumps mostly 3 to 6 inches high and 1 to several feet in diameter; joints bluish-green, orbiculate to broadly obovate, 2 to 4 inches long, 1 or usually 2 to 4 inches broad, about ⅜ inch thick, glabrous, not readily detached; spines distributed variously on the joints in the varieties, white to brown or reddish-brown or in age gray, about 6 to 10 per areole, mostly deflexed, straight or curving slightly downward, the longer ones 1 to 3¼ inches long, basally 1/96 to 1/48 inch in diameter, circular to elliptic in cross section, not strongly barbed; flower 1¾ to 3¼ inches in diameter; petaloid perianth parts yellow or occasionally pale or tinged with pink or rarely red (in some varieties), cuneate-obovate, 1 to 1⅝ inches long, ½ to ¾ inch broad, truncate to rounded; fruit dull tan or brown, dry at maturity, spiny (with barbed spines) over the entire surface, obovoid, ¾ to 1½ inches long, ½ to 1 inch in diameter or in sterile fruits parasitized by insect larvae larger, with the apical cup shallow, deciduous two or three months after flowering; seed light tan to nearly white, ⅛ to ¼ inch long.

Opuntia polyacantha is of lower stature than the related species, *Opuntia erinacea,* and the younger plants in particular are less compact. Older plants sometimes become semiprostrate from the weight of snow, and they root along the lower edges of the joints.

17A. Var. **rufispina**
(Engelm.) L. Benson, **comb. nov.** (p. 20)
(Table 4)

DISTRIBUTION: Sandy soils in the deserts and woodlands at 4,000 to 7,000 or sometimes 10,000 feet elevation. Juniper-Pinyon Woodland; Sagebrush Desert; Chihuahuan Desert. Northern Arizona from northeastern Mohave County to northern Apache County. California in the eastern

TABLE 4. CHARACTERS OF THE VARIETIES OF OPUNTIA POLYACANTHA

	A. Var. **rufispina**	B. Var. **juniperina**	C. Var. **trichophora**
Distribution of spines on the joint	In all the areoles.	In only the upper areoles.	In all the areoles.
Characters of the spines in the lower areoles	Rigid, straight, short, ½ to 1¼ inches long.	(None).	(Especially on basal joints) threadlike, flexible, curving or undulating, white or pale gray, elongate, 1½ to 3 inches long.
Length of longest spines in the upper areoles	Usually 1¾ to 3¼ inches.	1 to 1½ inches or (New Mexico) 2 inches.	1 or sometimes 1½ inches.
Color of larger spines	White, gray, or sometimes reddish-brown.	Reddish-brown.	Reddish-brown or pale gray.
Basal diameter of thickest spines	1/24 to 1/36 inch.	Up to 1/48 inch.	1/36 inch.
Fruits	Densely spiny.	With few spines, these deciduous.	With few to many rather weak spines.
Geographical distribution	Northern Arizona from northeastern Mohave County to northern Apache County. East of the southern Sierra Nevada, California, to westernmost Colorado, New Mexico, and western Texas.	Arizona in northeastern Apache County. Easternmost Utah and southernmost Colorado to northwestern New Mexico.	Northeastern Arizona in the Grand Canyon area and in Navajo County. Southwestern Colorado to northern New Mexico, western Oklahoma, and western Texas.

Fig. 1.48. *Opuntia polyacantha* var. *rufispina*. Joint and maturing (dry) fruit. (Photograph by Robert H. Peebles.)

edge of the Sierra Nevada and in the desert mountains southward to the New York Mountains; almost throughout the higher parts of Nevada and (less commonly) Utah; Wyoming near Green River; western Colorado. New Mexico from San Juan and Rio Arriba counties southward; Texas from El Paso County eastward to Jeff Davis County.

SYNONYMY: Opuntia rutila Nutt. Type locality: Colorado [Green] River about lat. 42° N. The name has been misapplied in several ways. *Opuntia missouriensis* DC. var. *rufispina* Engelm. & Bigelow. Type locality: Head of the Pecos River, New Mexico.

17B. Var. **juniperina**
      (Engelm.) L. Benson, **comb. nov.** (p. 20)
      (Table 4)

DISTRIBUTION: Sandy soils of flats, washes, and hillsides in the desert or grass-

land at 4,700 to 7,400 feet elevation. Sagebrush Desert and Plains Grassland or open grassy flats in the Juniper-Pinyon Woodland. Arizona in the Carrizo Mountains in northeastern Apache County. Near the Green River from southern Wyoming to the eastern edge of Utah and thence to western and middle Colorado and northwestern New Mexico.

SYNONYMY: Opuntia sphaerocarpa Engelm. & Bigelow. Type locality: Sandia Mountains, near Albuquerque, New Mexico. *Opuntia juniperina* Britton & Rose. *Opuntia erinacea* Engelm. & Bigelow var. *juniperina* W. T. Marshall. Type locality: Cedar Hill, San Juan County, New Mexico.

17C. Var. *trichophora*
      (Engelm. & Bigelow) Coulter
      (Table 4)

DISTRIBUTION: Sandy soils of plains, hills, or canyonsides in woodland at 5,000 or commonly about 8,000 feet elevation. Mostly Juniper-Pinyon Woodland. Arizona in eastern Coconino County and in Navajo County. Southwestern and southcentral Colorado; New Mexico, almost throughout the higher areas; Oklahoma in Cimarron County; Texas in the Panhandle and from Hudspeth County to Brewster County.

SYNONYMY: Opuntia missouriensis DC. var. *trichophora* Engelm. & Bigelow. *Opuntia polyacantha* var. *trichophora* Coulter. *Opuntia trichophora* Britton & Rose. Type locality: Santa Fe Creek, New Mexico.

18. *Opuntia fragilis*
     Nutt.
     Little prickly pear (Table 5)

Mat-forming plant, usually 2 to 4 inches high, the clumps often 1 foot or more in diameter; joints bluish-green, flattened-obovoid or -ovoid to elliptic or orbiculate, ⅔ to 1¾ inches long, ½ to 1 or 1¼ or even 1½ inches broad, ¾ inch thick, at least one-half as thick as broad, sometimes a few almost as thick as broad, readily detached, often clinging to animals by the barbed spines; spines usually on most of the joint but rarely none, the longest ones in the upper areoles, white or pale gray, 1 to 6 or 9 per areole, spreading in various directions, straight, the longer ones ½ to 1 or 1¼ inches long, basally up to 1/32 inch in diameter, from nearly circu-

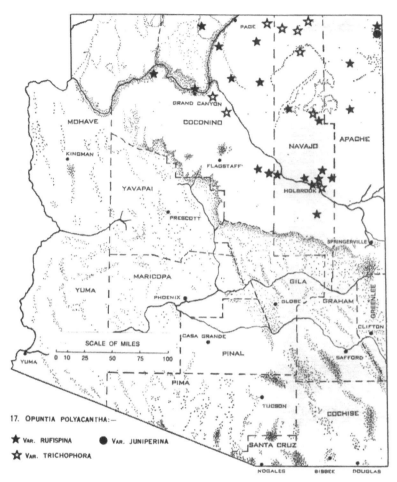

17. OPUNTIA POLYACANTHA:—

★ VAR. RUFISPINA   ● VAR. JUNIPERINA
☆ VAR. TRICHOPHORA

Fig. 1.49. The documented distribution of *Opuntia polyacantha*, according to its varieties.

lar to elliptic in cross section, strongly barbed; flower 1½ to 2 inches in diameter; petaloid perianth parts yellow or greenish or reported to be sometimes magenta, cuneate or cuneate-obovate, ⅝ to 1 inch long, ½ to ¾ inch broad, truncate to rounded, entire; fruit green or reddish-green before maturity, dry and tan at maturity, spiny or sometimes spineless in vegetatively spineless plants, obovoid, ½ to ⅝ inch long, ⅜ to ½ inch in diameter or larger in sterile fruits parasitized by insect larvae, with the apical cup deep, maturing and drying two or three months after flowering; seeds bone-color, ¼ inch long.

18A. Var. *fragilis*
     Little prickly pear (Table 5)

DISTRIBUTION: Sandy, gravelly, or rocky soils of valleys, hills, or mountainsides mostly in the desert from sea level to 2,000 feet elevation northward, but in Arizona mostly 3,000 to 5,000 or 8,000 feet elevation. Sagebrush Desert but also sparingly in the Pacific Forest, the Rocky Mountain Montane Forest, the Juniper-Pinyon Woodland, the Great Plains Grassland, and (rarely) the Prairie. Arizona from eastern Coconino County to Apache County. Southern and inland British Columbia to western Manitoba. Puget Sound, Washington; Columbia Basin and eastward through the Rocky Mountain System, the Great Plains, and the Great Lakes region as far as southern Michigan and Illinois and southward and southeastward to northeastern California, northeastern Nevada, northern New Mexico, Kansas, and the Oklahoma and Texas panhandles.

Hybrids of this species are common in Washington, Oregon, Idaho, eastern Utah, and western Colorado.

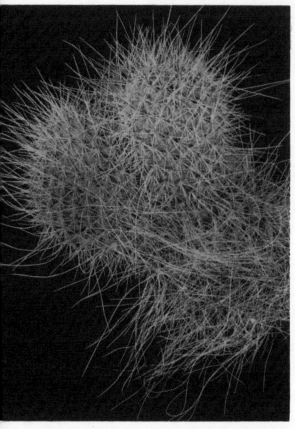

Fig. 1.50. *Opuntia polyacantha* var. *trichophora.* The related grizzly bear cactus, *Opuntia erinacea* var. *ursina,* has similar but flattened threadlike spines from the lower portions of especially the lower joints of the stem. (Photograph by Robert H. Peebles.)

SYNONYMY: Cactus fragilis Nutt. *Opuntia fragilis* Haw. *Tunas fragilis* Nieuwl and Lundell. Type locality: northern Great Plains. *Opuntia Schweriniana* K. Schuman. *Opuntia polyacantha* Haw. var. *Schweriniana* Backeberg. Type locality: Sapinero, Colorado. The name has been misapplied to several species. *Opuntia fragilis* (Nutt.) Haw. var. *caepitosa* L. H. Bailey. Type locality: "Colo." *Opuntia fragilis* (Nutt.) Haw. var. *tuberiformis* L. H. Bailey. Type locality: "Colo." *Opuntia fragilis* (Nutt.) Haw. var. *parviconspicua* Backeberg. Type locality: "Hab.?" *Opuntia fragilis* (Nutt.) Haw. var. *denudata* Wiegand & Backeberg. Type locality: "USA (Utah, im Utah-Bassin westlich von Steptoe Valley)." Steptoe Valley is in Nevada.

18B. Var. *brachyarthra*
      (Engelm. & Bigelow) Coulter
      (Table 5)

DISTRIBUTION: Sandy or gravelly soils of hills, plains, valleys, and mountainsides at mostly 4,500 to 8,000 feet elevation. Juniper-Pinyon Woodland and Great Plains Grassland. Arizona (rare) from northeasternmost Mohave County to northwestern Navajo County and the Grand Canyon region. Western Colorado; northwestern New Mexico.

SYNONYMY: Opuntia brachyarthra Engelm. & Bigelow. *Opuntia fragilis* (Nutt.) Haworth var. *brachyarthra* Coulter. Type locality: Inscription Rock near Zuñi, New Mexico.

19. *Opuntia erinacea*
      Engelm. & Bigelow

Clumps mostly 6 inches to 1 foot high and 1 yard or more in diameter; joints bluish-green, elongate, elliptic-oblong to obovate-oblong, 2 to 8, commonly 4 to 6, inches

TABLE 5. CHARACTERS OF THE VARIETIES OF OPUNTIA FRAGILIS

	A. Var. **fragilis**	B. Var. **brachyarthra**
Larger terminal joints	Elliptic to obovate or orbiculate, ¾ to 1½ or 1¾ inches long, ½ to 1 or 1¼ inches broad.	Elliptic to obovate (or narrowly either one) or orbiculate, 1 or 1½ to 1¾ or 2⅞ inches long, about 1 or rarely 1¼ or even 1½ inches broad.
Longest spines	½ to ⅝ or sometimes ¾ or rarely 1 inch long (or rarely the plant spineless), gray, tan, brown, or somewhat reddish, rigid.	¾ or 1 to 1¼ inches long, red or reddish-brown, rigid.
Geographical distribution	Northern, mostly northeastern, Arizona. British Columbia to the Great Lakes region, northeastern California, northwestern New Mexico, and the Texas Panhandle.	Northern (extreme northcentral) Arizona. Western Colorado and northwestern New Mexico.

Fig. 1.51. Little prickly pear, *Opuntia fragilis*. (Photograph by Robert H. Peebles.)

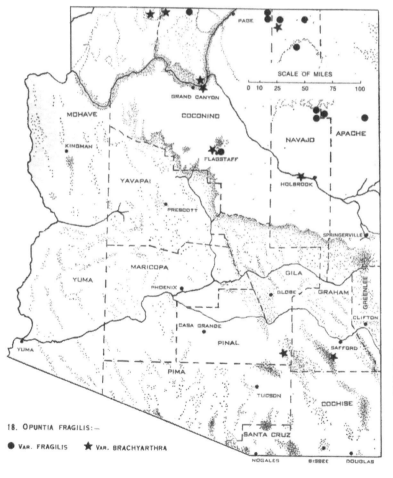

18. OPUNTIA FRAGILIS:—

● VAR. FRAGILIS   ★ VAR. BRACHYARTHRA

Fig. 1.52. The documented distribution of *Opuntia fragilis*, according to its varieties.

long, 2½ to 3½ inches broad, less than ½ inch thick, not more than one-fourth as thick as broad; spines occurring over the whole joint or (in var. *utahensis*) in only the upper areoles, white or pale gray, 4 to 7 or 9 per areole, turned downward, somewhat but not strongly twisted, flexible and curving irregularly, the longer ones usually about 2 but sometimes ½ or up to 4 inches long, basally about 1/48 inch broad, almost filiform, but at least some basally clearly flattened; flower about 1¾ to 3⅝ inches in diameter; petaloid perianth parts rose to deep pink or yellow, broadly cuneate or cuneate-obovate, 1 to 1⅜ inches long, ½ to ⅝ inch broad, truncate or retuse, mucronate, somewhat undulate; fruit tan to brownish, dry at maturity, densely spiny (the spines usually in a ring at the apex, spreading, strongly barbed, obovoid-cylindroid), 1 to 1¼ inches long, ½ to ¾ inch in diameter, with the apical cup deep, deciduous; seeds bone-white, irregularly discoid, ⅙ to ¼ inch long.

19A. Var. *erinacea*
    Mojave prickly pear
    (Table 6)

DISTRIBUTION: Sandy or gravelly soils of valleys, low hills, or canyonsides in the desert or woodland at 1,500 to 5,000 or 7,500 feet elevation. Mojavean Desert or Juniper-Pinyon Woodland or rarely Sagebrush Desert or Colorado Desert. Arizona from Mohave County to Apache County. Southern California in and near the Mojave Desert; the southern half of Nevada; southern Utah; New Mexico in western Socorro County.

Hybrid swarms of *Opuntia erinacea* and *Opuntia phaeacantha* var. *major* (of the juicy-fruited group of prickly pears) occur in southern Utah and northern Arizona. In Coconino County, Arizona, a similar series involves *Opuntia littoralis* var. *Martiniana*. Numerous intergradations between this species and *Opuntia phaeacantha* var. *major* have been observed by the writer in Bridge Canyon and along Glen Canyon. See discussion under *Opuntia Nicholii*.

Distinction of the complex of varieties composing *Opuntia erinacea* from the complex composing *Opuntia polyacantha* is more or less but not strictly clear on the basis of the flattening of at least the basal portions of the larger spines in *Opuntia erinacea*. Examples of intergradation appear especially along the geographical zone of contact between var. *erinacea* and *Opuntia polyacantha* var. *rufispina*.

SYNONYMY: Opuntia erinacea Engelm. & Bigelow. *Opuntia hystricina* Engelm. & Bigelow var. *Bensonii* Backeberg, *nom. nov.* for *Opuntia erinacea,* but an illegitimate epithet. Type locality: Mojave River, California. *Opuntia xerocarpa* Griffiths. Type locality: 15 miles southeast of Kingman, Arizona.

Fig. 1.53. *Opuntia erinacea.* Joint with a flower and young fruits. The small white arrows point to spines which are *flattened,* as shown by their seeming variation in "width" as seen in the photograph. (Photograph by Robert H. Peebles.)

19B. Var. *ursina*
(Weber) Parish
Grizzly bear cactus (Table 6)

DISTRIBUTION: Rocky soils of hillsides in the desert chiefly at 4,000 to 5,500 feet elevation. Higher parts of the Mojavean Desert. Arizona from Mohave County to Apache County. California in and near the Mojave Desert; southern Nevada; Utah in Washington County.

The variety is prized in cultivation for the long flexible undulating deflexed spines developed at the bases of the lower joints.

Similar spines occur in *Opuntia polyacantha* var. *trichophora* of southwestern Colorado, northeastern Arizona, southern New Mexico, and westernmost Oklahoma. The spines of var. *trichophora* are nearly circular in cross section. The elongate and threadlike character of the spines of the lower joints of both taxa may represent *either* parallel evolution *or* retention of this character with loss or modification of others derived from a common ancestry.

SYNONYMY: Opuntia ursina Weber. *Opuntia erinacea* Engelm. & Bigelow var. *ursina* Parish. Type locality: Ord Mountains, Mojave Desert, California.

19C. Var. **utahensis**
(Engelm.) L. Benson, **comb. nov.** (p. 20)
(Table 6)

DISTRIBUTION: Sandy or gravelly soils of plains and mountainsides in woodlands at 3,000 or commonly 5,600 to 8,000 feet elevation. Juniper-Pinyon Woodland or the upper edge of the Sagebrush Desert. Northern edge of Arizona from Mohave County to Apache County. Southern Idaho to California along the eastern side of the southern Sierra Nevada, the middle levels of Nevada and Utah, western Colorado and Canyon City, and northern New Mexico.

In New Mexico the variety is represented chiefly by forms shading into *Opuntia polyacantha* var. *juniperina*. Intergradation of var. *utahensis* with var. *erinacea* and with *Opuntia polyacantha* var. *rufispina* is discussed under var. *erinacea*, p. 75. Variety *utahensis* intergrades commonly with other varieties of *Opuntia polyacantha*.

SYNONYMY: Opuntia sphaerocarpa Engelm. & Bigelow var. *utahensis* Engelm., not *Opuntia utahensis* J. A. Purpus. Type locality: Steptoe Valley, Nevada. *Opuntia rhodantha* K. Schumann. *Opuntia*

TABLE 6. CHARACTERS OF THE VARIETIES OF OPUNTIA ERINACEA

	A. Var. **erinacea**	B. Var. **ursina**	C. Var. **utahensis**	D. Var. **hystricina**
Joint shape	Elliptic- to obovate-oblong, commonly 4 to 5 inches long, 1 to 1½ or 2 inches broad.	As in var. **erinacea**	Broadly to narrowly obovate or elliptic, mostly 2 to 3½ inches long, 2 to 3 inches broad.	Broadly deltoid-obovate, 1¼ to 2 inches long, of equal breadth.
Distribution of spines on the joint	In all areoles, the apical longest, reduced gradually down the joint, usually recurved or turned downward, somewhat flexuous, the longest 1¼ or 2 or 3¾ inches long, 1/48 inch broad.	In all areoles, the longest ones above, but those in the lower areoles also long and more slender, being turned downward, curving, and remarkably flexuous, the longest 3 to 4 inches long, 1/96 inch broad.	In only the upper areoles, spreading or (if long) deflexed, straight, or sometimes recurving, not flexuous, 1 to 1½ or 2¼ inches long, usually about 1/32 inch broad.	In all areoles, even those low on the joint spreading, nearly straight, not flexous, the longest 2½ to 3¾ inches long, 1/48 to 1/24 inch broad.
Spines of the fruit	Abundant above.	Abundant on the entire fruit.	Above, few and short.	Abundant.
Geographical distribution	Northern Arizona. Californian deserts to southwestern Colorado (rare) and New Mexico (Socorro County, rare)	Mohave County, Arizona. Mojavean Desert from California to Washington County, Utah.	Northern Arizona. Southern Idaho to California (southern Sierra Nevada), western Colorado (rare), and northern and westcentral New Mexico.	Northern Arizona on the watershed of the Little Colorado River.

Fig. 1.54. Prickly pears of sandy areas in northern Arizona. *Above*, Mojave prickly pear, *Opuntia erinacea. Below, Opuntia erinacea* var. *utahensis.*

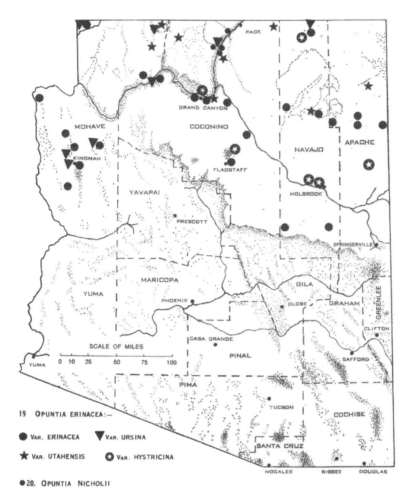

Fig. 1.55. The documented distribution of *Opuntia erinacea,* according to its varieties, and of *Opuntia Nicholii.*

19 OPUNTIA ERINACEA:—

● VAR. ERINACEA   ▼ VAR. URSINA

★ VAR. UTAHENSIS   ✪ VAR. HYSTRICINA

●20. OPUNTIA NICHOLII

Fig. 1.56. *Opuntia erinacea* var. *utahensis.* Branch with a flower bud, a flower, and a very young fruit, the spines restricted to the upper part of the joint. The small white arrow points to a flattened spine, as shown by the seemingly differing "width" of the spine at the base *(side view)* and farther out *(flat top view).* (Photograph by Robert H. Peebles.)

erinacea Engelm. & Bigelow var. *rhodantha* L. Benson. Type locality: Colorado. *Opuntia xanthostemma* K. Schumann. *Opuntia rhodantha* K. Schumann var. *xanthostemma* Rehder. *Opuntia erinacea* Engelm. & Bigelow var. *xanthostemma* L. Benson. Type locality: Grand Mesa, Colorado. *Opuntia rhodantha* K. Schumann var. *spinosior* Boissevain. No type specimen designated.

19D. Var. *hystricina*
       (Engelm. & Bigelow) L. Benson
       Porcupine prickly pear (Table 6)

DISTRIBUTION: Sandy plains, hills, and washes in deserts, grasslands, or woodlands at 5,000 to 7,300 feet elevation. Sagebrush Desert, Juniper-Pinyon Woodland, or Great Plains Grassland. Northern Arizona on the Little Colorado watershed.

The variety has been misinterpreted by all authors, having been confused commonly with long-spined forms of *Opuntia polyacantha* var. *rufispina*.

Some specimens from west-central New Mexico as far east as Albuquerque have characters of this variety.

SYNONYMY: Opuntia hystricina Engelm. & Bigelow. *Opuntia erinacea* Engelm. & Bigelow var. *hystricina* (Engelm. & Bigelow) L. Benson. Type locality: Little Colorado River in Arizona.

20. *Opuntia Nicholii*
     L. Benson
     Navajo bridge prickly pear

Clumps about 6 to 9 or 12 inches high, mostly 3 to 20 feet in diameter; joints bluish-green, narrowly obovate to sometimes obovate, 4 or commonly 5 to 8 inches long, 2 or mostly 3 to 5 inches broad, about ½ inch or more thick, not more than one-eighth as thick as broad; spines conspicuous, growing from all the areoles, those of the lower ones sometimes shorter, white or very pale gray, often reddish before maturity, about 4 to 7 per areole, turned downward, somewhat flexible, markedly twisting and curving in various directions, the longer ones (upper areoles) 3 to 5 inches long, these basally 1/24 to 1/16 inch broad, flattened, not barbed; flower 2½ to 3 inches in diameter; petaloid perianth parts magenta to rose or rarely yellow, cuneate-obovate, 1⅜ to 1⅝ inches long, 1 to 1¼ inches broad, mucronate, slightly undulate; fruit tan to brownish, dry at maturity, with a small number of strongly-barbed horizontally-spreading or deflexed spines mostly around the apices (the minute barbs at the

apex of the spine), 1 to 1⅜ inches long, ¾ to ⅞ inch in diameter, with a shallow apical cup, deciduous two or three months after flowering; seed bone-white, irregular, 5/16 inch long.

DISTRIBUTION: Gravelly soils of flats and low ridges in the desert at 4,000 to 5,000 feet elevation. Navajoan Desert. Arizona in Coconino County from the Utah Border near the Colorado River to the broad shelves between the river and Vermilion and Echo Cliffs. Utah in San Juan County along and near Glen Canyon of the Colorado River from Ticaboo Canyon to Bridge Canyon.

The plants resemble *Opuntia erinacea,* but they are much larger. The long, white, deflexed spines curve in bizarre fashion.

The species combines the following characters of species not closely related to each other:

Opuntia erinacea: Spines somewhat flattened, white or very pale gray, turned downward, markedly twisted, relatively flexible; glochids small; flowers nearly always magenta to rose; fruits dry at maturity, with strongly barbed, horizontally spreading spines on the upper parts; seeds large.

Opuntia phaeacantha var. *major* (Series *Opuntiae*): Joints of similar form but much larger than in *Opuntia erinacea;* spines of coarse texture, stout and stiff, in a few individuals mostly in the apical areoles, reddish when young; flowers yellow or largely so, in rare instances reddish purple.

As prickly pear populations are constituted, the plants growing at 3,800 to 4,200 feet elevation south and east of the Navajo Bridge are nearly uniform in their characters. However, despite the degree of stability of the characters distinguishing *Opuntia Nicholii* from its relatives, there is some variation from individual to individual. This shows clearly that the plants are not a single clone formed by growth of detached fragments of one original individual.

When *Opuntia Nicholii* was named and described in 1950, the writer postulated a hybrid origin in the remote past from *Opuntia erinacea* and *Opuntia phaeacantha* var. *major* followed by natural selection of a population adapted to the local habitat. This possibility was suggested by hybridizing of these species observed at Springdale, Utah, and elsewhere. Since then both species and a hybrid swarm appearing to have resulted from their interbreeding have been found about ten miles southeast of the Navajo Bridge at about 1,000 feet higher elevation. In some other hybrid swarms of the two species, plants similar to *Opuntia Nicholii* were not found, but in this area some individuals were identical with it while others graded into *Opuntia erinacea* or into *Opuntia phaeacantha* var. *major.*

The status of *Opuntia Nicholii* is open to question. It differs from both the probable parental

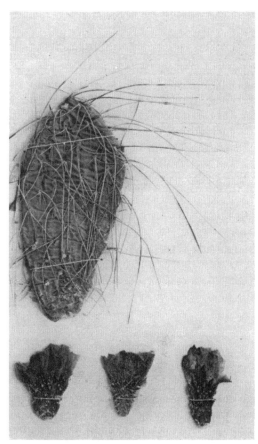

Fig. 1.57. Navajo Bridge Prickly Pear, *Opuntia Nicholii*. *Left,* greatly enlarged dry fruits and seeds. *Right,* stem with the characteristic very long, stout spines; flowers (sewed on an herbarium sheet).

species through a large area at a particular elevation, and its rather narrow range of gene combinations is well sorted out. *Opuntia Nicholii* is really an incipient species and not clearly differentiated. Reduction to varietal status may be a reasonable solution to the problem, but it is difficult to know under which species subordinate status should be accorded. Almost as much may be said for one as for the other. However, if there is any choice, it may be in favor of *Opuntia erinacea.*

SYNONYMY: Opuntia Nicholii L. Benson. *Opuntia hystricina* Engelm. & Bigelow var. *Nicholii* Backeberg. Type locality: Navajo Bridge, Colorado River, Coconino County, Arizona.

Series 2
BASILARES
Glochid Cacti

Mats or clumps or sometimes shrubs several feet high. Joints in the species occurring in the United States with a bluish bloom, usually finely hairy; rarely, except in Mexican species, with spines but always with many glochids. Fruit either dry or, in species not occurring in Arizona, fleshy. Seeds large, $\frac{1}{8}$ to $\frac{1}{4}$ inch in diameter, smooth or rough and irregular.

DISTRIBUTION: Two species in the United States. Others native in Mexico.

21. *Opuntia basilaris*
    Engelm. & Bigelow
    Beavertail cactus

Clumps usually 6 to 12 inches high, 1 to 6 feet in diameter; joints blue-green, in cold weather also irregularly purplish, usually minutely and densely hairy and ashy blue-green and velvety but sometimes not hairy, obovate or sometimes circular to narrowly elongate

or spathulate, 2 to 6 or 13 inches long, 1 or
1½ to 4 or 6 inches broad, about ½ inch
thick; spines none, except in var. *Treleasei;*
glochids brown or reddish-brown, about ⅛
inch long, troublesome; flower about 2 to 3
inches in diameter, 2 to 3 inches long; petaloid
perianth parts cerese or (in one variety) yellow, obovate or cuneate, 1 to 1½ inches long,
½ to 1 inch broad, truncate to rounded, cuspidate, undulate-crenate; fruit green, changing to tan or gray, dry at maturity, not spiny,
except with a few spines in var. *Treleasei,* 1 to
1¼ inches long, ⅝ to ⅞ inches in diameter,
with an apical cup, deciduous within three or
four months after flowering; seed bone-white
or grayish, smooth or sometimes rough and
irregular, and then the margin conspicuous
or corky, about ¼ inch long.

21A. Var. *basilaris*
      Beavertail cactus
      (Table 7)

DISTRIBUTION: Sandy, gravelly, or
rocky soils of valleys, washes, or canyons in
the desert from sea level to 4,000 feet elevation but sometimes up to 5,000 or even 9,000
feet. Mojavean and Colorado Deserts; the
edge of the Sagebrush Desert; California Oak
Woodland (rare); California Chaparral; Juniper-Pinyon Woodland; Sierran Montane Forest (rare). Arizona from Mohave County to
Coconino County (at low altitudes along the
Colorado River) and to Yavapai, western
Yuma, and northwestern Maricopa counties.
California, occasional on the coastal sides the
southern mountains just west of the deserts
and common in the Mojave and Colorado
deserts and rare on the east side of the southern Sierra Nevada; southern Nevada; Utah in
Washington and San Juan counties (rare).
Mexico in northern Sonora.

SYNONYMY: Opuntia basilaris Engelm. & Bigelow. Type locality: Cactus Pass, Mohave County,
Arizona. *Opuntia basilaris* Engelm. & Bigelow var.
ramosa Parish. Type locality: San Bernardino
Mountains, California. *Opuntia basilaris* Engelm. &
Bigelow var. *cordata* F. Fobe. No specimen or
locality mentioned. *Opuntia intricata* Griffiths. Type
locality: near San Bernardino, California. *Opuntia
humistrata* Griffiths. *Opuntia basilaris* Engelm. &
Bigelow var. *humistrata* W. T. Marshall. Type locality: above San Bernardino, California. *Opuntia
whitneyana* Baxter. *Opuntia basilaris* Engelm. &
Bigelow subsp. *whitneyana* Munz. Type locality:
Alabama Hills, Inyo County, California. *Opuntia
whitneyana* Baxter var. *albiflora* Baxter, not *Opuntia albiflora* K. Schumann, not *Opuntia basilaris* var.
albiflorus Walton. Type locality: near Mount Whit-
ney, California.

TABLE 7. CHARACTERS OF THE VARIETIES OF OPUNTIA BASILARIS

	A. Var. **basilaris**	B. Var. **longiareolata**	C. Var. **aurea**	D. Var. **Treleasei**
Joint shape	Obovate or sometimes orbiculate.	Spathulate.	Elliptic to obovate.	Narrowly elliptic or obovate.
Joint size	2 to 6 or 13 inches long, 1½ to 4 or 6 inches broad.	4 to 5 inches long, about 2 inches broad.	2 to 4 inches long, 1¼ to 2 or 2½ inches broad.	3 to 10 inches long, 2 to 4 inches broad.
Areole shape & size	Circular, 1/16 to ⅛ inch in diameter.	Elongate, ⅛ inch long, 1/24 inch broad.	As in var. **basilaris**	Circular, ⅛ inch broad.
Flower color	Cerese.	Cerese.	Yellow.	Cerese.
Seed margin	Inconspicuous.	Unknown.	Large and irregular.	Inconspicuous.
Geographical distribution	Arizona from Mohave County to Yavapai, west-ern Maricopa, and western Yuma counties. Southern California in and near the deserts to southernmost Utah. Mexico in northernmost Sonora.	Arizona at Granite Rapids, Grand Canyon.	Northern edge of Arizona. Southern edge of Utah.	Northwesternmost Ari-zona. California in the San Joaquin Valley in Kern County from northeast to southeast of Bakersfield and in the Turtle Mountains (Mojave Desert) and perhaps the San Gabriel Mountains.

Fig. 1.58. Beavertail cactus, *Opuntia basilaris*, in flower. (Photographs: *above*, by Walter S. Phillips; *below*, by Robert A. Darrow.)

Fig. 1.59. Beavertail cactus, *Opuntia basilaris*. The characteristic spineless joints; the areoles with numerous troublesome glochids. (Photograph by Robert H. Peebles.)

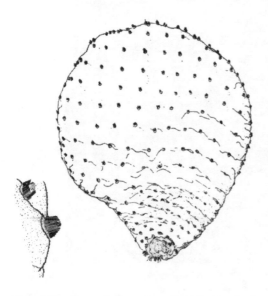

Fig. 1.60. Beavertail cactus, *Opuntia basilaris*. The joint is densely covered with minute hairs, as shown in the enlargement of a portion of the stem and of an areole, which shows the glochids.

21B. Var. *longiareolata*
       (Clover & Jotter) L. Benson
       Grand Canyon beavertail (Table 7)

DISTRIBUTION: Rocky bases of talus slopes in the desert at about 2,000 feet elevation. Mojavean Desert. Arizona at Granite Rapids, Grand Canyon, Coconino County.

The validity of this variety is dubious. It is maintained pending field study to determine the existence of a population or of just an individual. The plant is known only from the type collection.

SYNONYMY: Opuntia longiareolata Clover & Jotter. *Opuntia basilaris* Engelm. & Bigel. var. *longiareolata* L. Benson. Type locality: Granite Rapids, Grand Canyon, Coconino County, Arizona.

21C. Var. *aurea*
       (Baxter) W. T. Marshall
       Yellow beavertail (Table 7)

DISTRIBUTION: Sand of flats, dunes, and valleys in woodland areas at 4,000 to 7,000 feet elevation. Juniper-Pinyon Woodland and the upper edge of the Sagebrush Desert. Arizona along the Utah boundary

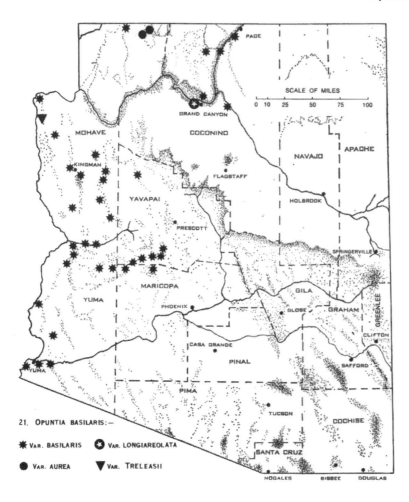

Fig. 1.61. The documented distribution of *Opuntia basilaris*, according to its varieties.

21. OPUNTIA BASILARIS:—

✳ VAR. BASILARIS   ✪ VAR. LONGIAREOLATA

● VAR. AUREA   ▼ VAR. TRELEASII

north of Pipe Spring, Mohave County. Utah in southeastern Washington County and in southern Kane County.

SYNONYMY: Opuntia aurea Baxter. *Opuntia basilaris* Engelm. & Bigelow var. *aurea* W. T. Marshall. *Opuntia lubrica* Griffiths var. *aurea* Backeberg. Type locality: north of Pipe Spring, Mohave County, Arizona.

21D. Var. *Treleasei*
    (Coulter) Toumey
    Kern cactus (Table 7)

DISTRIBUTION: Sandy soils of flats and low hills in grasslands at 400 to 1,000 feet elevation. Northwestern Arizona near the Colorado River. California in the San Joaquin Valley in Kern County and in the Turtle Mountains, eastern Mojave Desert.

SYNONYMY: Opuntia Treleasii [*Treleasei*] Coulter. *Opuntia basilaris* Engelm. & Bigelow var. *Treleasei* Toumey. Type locality: Caliente, Tehachapi Mountains, California. *Opuntia Treleasei* Coulter var. *Kernii* Griffiths & Hare. Type locality: East Bakersfield, California.

Series 3
OPUNTIAE
Prickly Pears

Shrubs, trees, or matted plants. Joints green or bluish, usually but not necessarily without hair and spiny. Fruit (if fertile) purple, lavender, or red; fleshy, juicy, spineless but usually with effective glochids. Seeds small, 1/8 to 3/16 or 1/4 inch in diameter, smooth and regular.

DISTRIBUTION: Eleven native and three introduced species in the United States and

Canada from California to Colorado and Texas and eastward to Ontario, Massachusetts, and Florida. Five in Arizona. Many in Latin America.

KEY TO THE SPECIES

1. Spines present.
  2. Spines needlelike, nearly as thick as broad, elliptic to nearly circular in cross section (but sometimes an occasional spine flattened basally), only 1 to 3 per areole, not yellow.
    3. Plants low and mat-forming, usually prostrate, the joints in series of only a few (usually 3 to 5); joints usually 2 to 3 or 4 inches long, 1½ to 3 inches broad; spines white or gray; Utah and northern and eastern Arizona to the Great Plains.
        22. *Opuntia macrorhiza*, page 86
    3. Plants arising the height of several joints, mostly 2 to 3 or 7 feet high but the joints sometimes forming long series along the ground; joints usually 4 to 7 or rarely 12 inches long, usually 3 to 6 or 8 inches broad; spines tan, brown, pink, gray, black, reddish-brown, or sometimes white or partly yellow.
      4. Spines in the upper areoles of the joint 3 to 7 or 11, but often some of them small and little exceeding the glochids, brown, tan, pink, gray, or partly yellow; joints green during favorable seasons though sometimes lavender to reddish-purple in cold or dry weather; petaloid perianth segments commonly yellow or magenta.
          23. *Opuntia litoralis*, page 89
      4. Spine(s) in the upper areoles of the joint 1 or sometimes 2; joints strongly tinted with lavender or reddish-purple at all seasons; petaloid perianth parts pale yellow with red bases.
          24. *Opuntia violacea*, page 91
  2. Spines (at least some of the larger but not necessarily the smaller ones) *flattened at least at the bases (i.e.,* markedly broader than thick and narrowly elliptic in cross section), tapering, usually 3 or more per areole. (An exceptional species has narrow, sometimes only slightly flattened spines; these are yellow but becoming dirty dark gray in age or on old dried specimens.)
    3′. Spines *not all* pointing downward, spreading in various directions, no single lower spine markedly longer than the others in the areole, brown, brownish-red, pale gray, or white, never yellow; fruit obovoid to elongate; plant with no trunk.
          25. *Opuntia phaeacantha* (including "Opuntia Engelmannii"), page 95
    3′. Spines *all* (1 to 6 or 8 per areole) pointing downward, one lower spine markedly longer than the others, straight; spines clearly yellow, changing (often through white first)

to black in age or in dried specimens; fruit spheroidal or nearly so; plant 3 to 6 feet high, with a definite short trunk.
          26. *Opuntia chorotica*, page 103
1. Spines none.
  2′. Joints broadly obovate to obicular, 5 to 8 inches broad, lavender or purple at all seasons but this intensified during cold or drought; erect or suberect plants; southeastern Arizona to Texas west of the Pecos River.   24C. *Opuntia violacea* var. *santa-rita*, page 92
  2′. Joints narrowly obovate, 4½ to 6 inches broad, green except during extreme cold or drought; sprawling plants growing only on rocky ledges of mountains; southeastern Arizona.   25A. *Opuntia phaeacantha* var. *laevis*, page 97

22. *Opuntia macrorhiza*
Engelm.
Plains prickly pear

Clump-forming prickly pear, usually 3 to 5 inches high, 1 to 6 feet in diameter; the main root(s) usually tuberous, the other (adventitious) roots along the joints usually fibrous; joints bluish-green, orbiculate to obovate, 2 to 4 inches long, 2 to 2½ or 3 inches broad; spines mostly from the uppermost areoles, white or gray or rarely brownish or reddish-brown, 1 to 6 per areole, mostly turned downward, straight or slightly curving, the longer ones 1½ to 2¼ inches long, slender, basally 1/96 or 1/48 inch in diameter, needlelike, nearly circular or elliptic in cross section, not barbed; flower 2 to 2½ inches in diameter and of equal length; petaloid perianth parts yellow or tinged basally with some red, cuneate-obovate, 1 to 1½ inches long, ½ to 1 inch broad, rounded, entire; fruit purple or reddish-purple, fleshy at maturity, with some glochids, obovoid, 1 to 1½ inches long, 1 to 1⅛ inches in diameter, with a shallow apical cup, persistent for several months; seeds pale tan or gray, irregular but basically more or less discoid, about 3/16 inch in diameter.

22A. Var. *macrorhiza*

Stems moderately bluish; joints usually 2½ or 3 to 4 inches long; spines slender, about 1/48 inch in diameter; flowers yellow or the centers reddish.

DISTRIBUTION: Sandy, gravelly, or rocky soils of plains or valleys or less commonly of hillsides at 2,000 to 7,000 or 8,000

Fig. 1.62. Plains prickly pear, *Opuntia macrorhiza. A,* Plant in fruit. *B,* joint with a flower. *C,* Plant dug up to show the fleshy, enlarged tap root. (Photos by Robert H. Peebles.)

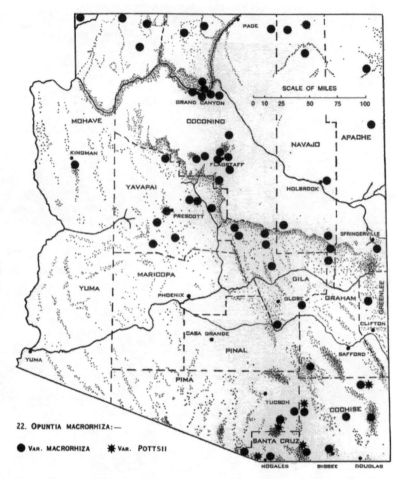

Fig. 1.63. The documented distribution of *Opuntia macrorhiza*, according to its varieties.

22. OPUNTIA MACRORHIZA: ——

● VAR. MACRORHIZA   ✳ VAR. POTTSII

feet elevation. Great Plains Grassland, Juniper-Pinyon Woodland, and the lower edge of the Rocky Mountain Montane Forest; rare in the Prairie or the edges of the Deciduous Forests. High plains of Arizona north and east of the Mogollon Rim. California on Clark Mountain, San Bernardino County; Utah (rare); eastward to the Great Plains Region from Wyoming (Cheyenne) and southwestern South Dakota to Nebraska, western Missouri, western Arkansas, and middle and southern Texas; Minnesota; Wisconsin; Michigan; Iowa; Illinois; Ohio (introduced east of Cincinnati); Louisiana (Cameron). Rare in the Middle West.

Especially in Arizona, New Mexico, and Texas this species shades into *Opuntia phaeacantha*. Eastward it intergrades somewhat with *Opuntia compressa,* and distinction from that species is not wholly clear. Plants with intermediate combina-

tions of characters have received various scientific names on the supposition that they were new species.

SYNONYMY: *Opuntia macrorhiza* Engelm. *Opuntia mesacantha* Raf. var. *macrorhiza* Coulter. *Opuntia compressa* (Salisb.) Macbr. var. *macrorhiza* L. Benson. Type locality: "Comanche Spring, New Braunfels, etc.," Texas. *Opuntia tortispina* Engelm. & Bigelow. Type locality: "On the Comanche plains, near the Canadian River." *Opuntia cymochila* Engelm. & Bigelow. *Opuntia mesacantha* Raf. var. *cymochila* Coulter. *Opuntia tortispina* Engelm. & Bigelow var. *cymochila* Backeberg. Type locality: Tucumcari hills, New Mexico. *Opuntia cymochila* Engelm. & Bigelow var. *montana* Engelm. & Bigelow. Type locality: Sandia mountains, near Albuquerque, New Mexico. *Opuntia stenochila* Engelm. & Bigelow. *Opuntia mesacantha* Raf. var. *stenochila* Coulter. Type locality: Zuñi, New Mexico. *Opuntia fusiformis* Engelm. & Bigelow. Type locality: "Kansas and Nebraska, in the regions of the Cross-Timbers, from the Canadian to the Big Bend of the Missouri." Deer Creek. *Opuntia mesacantha* Raf. var. *Greenei* Coulter. Type locality:

Golden, Colorado. *Opuntia mesacantha* Raf. var. *oplocarpa* Coulter. *Opuntia oplocarpa* Engelm. ex Coulter, *pro syn.* Type locality: Golden, Colorado. *Opuntia plumbea* Rose. Type locality: San Carlos, Arizona. *Opuntia utahensis* J. A. Purpus. Type locality: Piñon Valley, LaSal Mountains, Utah. *Opuntia Roseana* Mackensen. Type locality: San Antonio, Texas. *Opuntia leptocarpa* Mackensen. Type locality: San Antonio, Texas. *Opuntia Mackensenii* Rose. Type locality: Kerrville, Texas. The plants are intermediate between *Opuntia macrorhiza* and *Opuntia phaeacantha* var. *major*. *Opuntia seguina* C. Z. Nelson. Type locality: Seguin, Texas. *Opuntia MacAteei* Britton & Rose. Type locality: Rockport, Texas. *Opuntia Loomisii* Peebles. Type locality: Prescott, Arizona.

22B. Var. **Pottsii**
(Salm-Dyck) L. Benson **comb. nov.** (p. 20)

Stems strongly bluish, the joints 2 to 2½ inches long; spines very slender, about 1/96 inch in diameter; flowers reddish.

DISTRIBUTION: Sand or loam of plains and alluvial fans at 2,600 or 4,000 to 6,000 feet elevation. Grasslands and sometimes the Chihuahuan Desert. Arizona in southeastern Pima County and Cochise County. Southern New Mexico; Texas in the Panhandle and west of the Pecos River.

According to Margery S. Anthony, Amer. Midl. Nat. 55:243. 1956, the tuberous roots have milky juice.

SYNONYMY: Opuntia Pottsii Salm-Dyck. Type locality: near Chihuahua, Mexico. *Opuntia setispina* Engelm. Type locality: west of Chihuahua, Mexico. *Opuntia tenuispina* Engelm. & Bigelow. Type locality: near El Paso, Texas. *Opuntia filipendula* Engelm. Type locality: Rio Grande bottom near San Elisario. *Opuntia Ballii* Rose. Type locality: Pecos, Texas. *Opuntia delicata* Rose. Type locality: Calabasa[s], southeastern Arizona.

23. *Opuntia littoralis*
(Engelm.) Cockerell

Suberect or sprawling, usually 1 or 2 feet high and 2 to 4 feet in diameter; trunk none; joints green to noticeably or sometimes strongly glaucous, narrowly obovate or narrowly elliptic to broadly so or sometimes nearly orbiculate, 3 or 5 to 7 or 12 inches long, 2 or 3 to 4 or 5 inches broad; spines distributed over the entire joint or on only the upper part, brown, tan, pink, gray, or various combinations of these and yellow, 1 to 11 (or 0) per areole, spreading or some deflexed, usually straight, 1 to 2¼ inches long,

basally up to 1/24 inch in diameter, circular to elliptic in cross section, not barbed; flower 2 to 3 inches in diameter; petaloid perianth parts yellow with red or magenta bases or sometimes magenta or rose-purple, the largest obovate or obovate-cuneate, 1 to 1⅞ inches long, ⅝ to ⅞ inch broad, rounded and mucronate, nearly entire; fruit reddish to reddish-purple, fleshy at maturity, with only small glochids, obovoid or narrowly so, 1⅜ to 1⅝ inches long, about 1 to 1½ inches in diameter, with an apical cup, maturing after several months; seed light tan or gray, with the margin enclosing the embryo conspicuous and irregular, remarkably variable in size, ⅛ to ¼ inch in diameter.

23A. Var. *Martiniana*
(L. Benson) L. Benson

Joints moderately bluish, obovate to orbiculate, 4 or 5 to 7 inches long, 3 to 5½ inches broad; spines usually on most or all the joint, red-and-yellow to gray, 1 or 1½ to 1¾ inches long, 1 to 4 or 6 per areole; flower yellow or the center red.

DISTRIBUTION: Sandy or gravelly soils of valleys and mountainsides in forests or woodlands or the edge of the desert at 2,000 or 4,500 to 6,500 or 8,500 feet elevation. Lower edge of the Rocky Mountain Montane Forest; Juniper-Pinyon Woodland and Mojavean Desert. Arizona from westcentral Mohave County to the northern edge of the Kaibab Plateau in northern Coconino County and to northern Navajo County and to Yavapai County. Southern California in eastern San Bernardino County; southern Nevada; Utah along the Arizona border.

The variety hybridizes with *Opuntia erinacea*, and it shades into *Opuntia phaeacantha* and *Opuntia macrorhiza*.

SYNONYMY: Opuntia charlestonensis Clokey. *Opuntia phaeacantha* Engelm. var. *charlestonensis* Backeberg. Type locality: Charleston Mountains, Nevada. The type colony, studied in the field as well as in herbaria, shows, as do some other populations of var. *Martiniana*, a tendency toward inclusion of some characters of *Opuntia phaeacantha* var. *major*, which occurs nearby at lower elevations. *Opuntia macrocentra* Engelm. var. *Martiniana* L. Benson. *Opuntia littoralis* (Engelm.) Cockerell var. *Martiniana* L. Benson. Type locality: Hualpai Mountain, Mohave County, Arizona.

Fig. 1.64. *Opuntia macrorhiza* var. *Pottsii. A, B,* young joints with leaves and old ones without them; *B,* with a flower bud and a fruit; *C,* plant in cultivation. (Photographs by David Griffiths.)

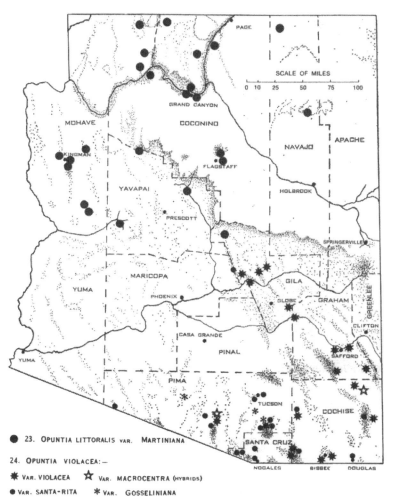

Fig. 1.65. The documented distribution of *Opuntia littoralis* var. *Martiniana* and of *Opuntia violacea*, according to its varieties.

● 23. OPUNTIA LITTORALIS var. MARTINIANA

24. OPUNTIA VIOLACEA: —

✳ VAR. VIOLACEA   ☆ VAR. MACROCENTRA (HYBRIDS)
● VAR. SANTA-RITA   ✱ VAR. GOSSELINIANA

24. *Opuntia violacea*
    Engelm.
    Purple prickly pear

Sprawling shrub usually 2 to 3 or 7 feet high, or sometimes treelike and then with a short trunk; joints green but at all seasons strongly tinted with or obscured by reddish-purple pigments, tending to be orbiculate, usually 4 to 8 inches long and of about equal breadth, about ⅜ inch thick; spines few or none, usually restricted to the upper part of the joint and often to the margin, in the Arizona varieties dark reddish-brown to almost black or sometimes pink, 1 (or in two varieties 2 to 3) per areole, spreading at right angles to the joint or vertically from the upper margin, straight or curving, somewhat flexible, 2 to 4 or 7 inches long, basally 1/72 to 1/24 inch in diameter, broadly elliptic to nearly circular in cross section, not barbed; flower about 2½ to 3½ inches in diameter; petaloid perianth parts few, yellow with bright red bases and lower middles, cuneate-obovate, 1 to 1½ inches long, ¾ to 1⅛ inches broad, rounded or truncate, nearly entire; fruit red or purplish-red, fleshy at maturity, smooth, 1 to 1½ inches long, ¾ inch in diameter, with an apical cup, persisting until about November; seeds light tan or gray, irregularly elliptic, 3/16 inch long.

The purplish to lavender color of the joints is characteristic of all the varieties at all seasons, but in times of drought or cold the color deepens, and it may become red. In the hot rainy season of July and August the joints become blue-green tinged with purple.

24A. Var. *violacea*
       (Table 8)

DISTRIBUTION: Sand or gravel or rocky soils of hills, mesas, washes, and canyons in grassland or along the upper edge of the desert at 3,000 to 5,000 feet elevation. Desert Grassland or the upper edge of the Arizona Desert. Southeastern Arizona from Gila County to Pima and Cochise counties. Southwestern New Mexico in Grant and Hidalgo counties.

SYNONYMY: Opuntia violacea Engelm. Type locality: near Solomon, Graham County, Arizona.

24B. Var. **Gosseliniana**
       (Weber) L. Benson, **comb. nov.** (p. 21)
       (Table 8)

DISTRIBUTION: Sandy or gravelly soils in grassland and the desert at usually 3,000 to 4,000 feet elevation. Desert Grassland or the upper edges of the Sonoran Deserts. Arizona in southern Pima County and the Huachuca Mountains in Cochise County. Mexico in Sonora.

SYNONYMY: Opuntia Gosseliniana Weber. Type locality: Sonora, Mexico.

24C. Var. **santa-rita**
       (Griffiths & Hare) L. Benson
       comb. nov. (p. 21)
       Purple prickly pear (Table 8)

DISTRIBUTION: Sandy or gravelly soils of plains and sometimes of canyons at 3,000 to 5,000 feet elevation. Desert Grassland or the edges of the Southwestern Oak Woodland or the Arizona and Chihuahuan deserts. Arizona in Gila County and from Pima County to Santa Cruz and Cochise counties. New Mexico from Hidalgo and Sierra counties to Otero County; Texas in Jeff Davis, Presidio, and Brewster counties. Mexico in northern Sonora.

This variety is strongly purple-to-lavender. It is cultivated because of its color and its spineless nearly circular joints. Sometimes it becomes very large and almost arborescent, having a trunk; forms of this type are popular.

The variety is not strongly resistant to drought, extra moisture, or overgrazing. Its spineless joints are vulnerable to rodents and to cattle.

SYNONYMY: Opuntia chlorotica Engelm. var. *santa-rita* Griffiths & Hare. *Opuntia santa-rita* Rose. *Opuntia Gosseliniana* Weber var. *santa-rita*. L. Benson. Type locality: Celero [Santa Rita] mountains, Arizona. *Opuntia Shreveana* C. Z. Nelson. Type locality: Tucson, Arizona.

24D. Var. **macrocentra**
       (Engelm.) L. Benson, **comb. nov.** (p. 21)
       Black-spined prickly pear (Table 8)

DISTRIBUTION: Sandy, gravelly, or other soils of plains, hills, and washes at 3,500 to 5,500 feet elevation. Mostly Desert

TABLE 8. CHARACTERS OF THE VARIETIES OF OPUNTIA VIOLACEA

	A. Var. **violacea**	B. Var. **Gosseliniana**	C. Var. **santa-rita**	D. Var. **macrocentra**
Joint shape and length	Obovate, 4 to 6 inches long.	Orbicular or broadly obovate, 5 to 7 inches long.	Orbicular, 6 to 8 inches long.	Mostly broadly obovate, 6 to 7 inches long.
Spines	Abundant near the upper margin of the joint, 1 to 3 per areole, dark reddish-brown, up to 1½ to 2½ inches long, 1/36 inch in diameter.	On both the upper and the flat sides of the joint, 1-2 per areole, light reddish-brown to pink or partly yellow, up to 1½ to 2½ inches long, 1/48 to 1/36 inch in diameter.	None or few, often 2-4 on the upper margin of the joint, 1 per areole, light reddish-brown to pink or rarely darker, up to 1½ to 2½ inches long, 1/72 to 1/32 inch in diameter.	Abundant on and just below the upper margin of the joint, 1 or 2 per areole, nearly black, commonly up to 3 to 7 inches long, 1/24 inch in diameter.
Geographical distribution	Southeastern Arizona from Gila to Pima and Cochise counties. Southwestern New Mexico in Grant and Hidalgo counties.	Southeastern Arizona in southern Pima County and in the Huachuca Mountains. Mexico in Sonora.	Arizona from Gila and Pima counties to Santa Cruz and Cochise counties. New Mexico from Hidalgo County to Sierra and Otero counties; Texas in Jeff Davis, Presidio, and Brewster counties.	Arizona plants approaching the variety are rare west of Tucson and from Benson eastward. Southern New Mexico in Dona Ana and Otero counties; Texas west of the Pecos River. Northern Mexico in Chihuahua.

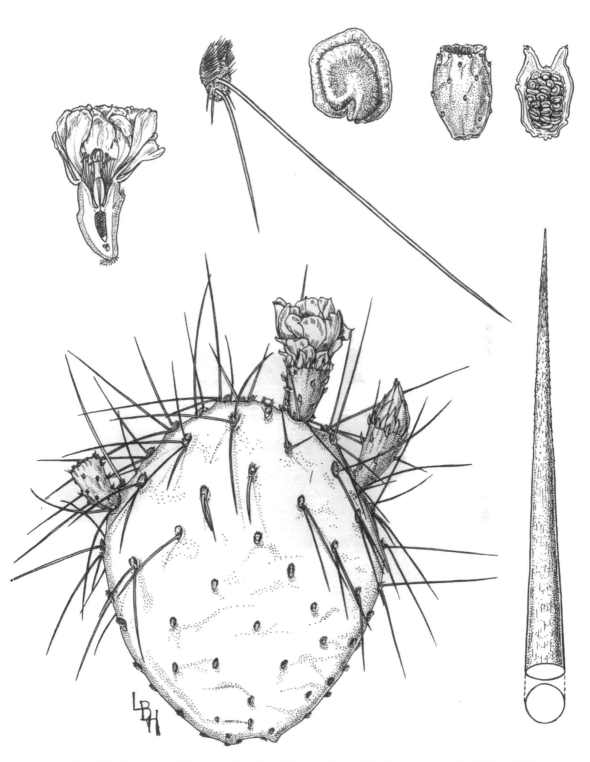

Fig. 1.66. Purple prickly pear, *Opuntia violacea*. *Upper left,* flower in longitudinal section. *Above,* spine cluster. *Upper right,* enlarged seed; fruit in side view and in longitudinal section. *Below,* joint with flower bud, flower, and young fruit; enlarged spine.

Fig. 1.67. Purple prickly pear, *Opuntia violacea*, in flower. *Above*, var. *violacea*. *Below*, var. *macrocentra*.

Fig. 1.68. Purple prickly pear, *Opuntia violacea* var. *santa-rita*. This is the spineless or nearly spineless variety, which tends to become a tree. (Photograph by David Griffiths.)

Grassland but also in the Southwestern Oak Woodland and the Arizona and Chihuahuan deserts. Arizona (forms approaching this variety, these being rare) west of Tucson and from Benson eastward. Southern New Mexico in Hidalgo (rare), Doña Ana, and Otero counties; Texas west of the Pecos River from El Paso County to Terrell County. Mexico in Sonora and Chihuahua.

SYNONYMY: Opuntia macrocentra Engelm. Type locality: El Paso, Texas.

25. *Opuntia phaeacantha*
    Engelm.

DISTRIBUTION: Prostrate or sprawling large prickly pears, the clumps 2 to 8 or 20 feet in diameter and 1 to 2 or 3 feet high, in some varieties occasionally with chains of several joints on edge along the ground; trunk none; joints bluish-green, in cold weather with some lavender-to-purple pigmentation, obovate or narrowly so or orbiculate, 4 to 6 or 10 or rarely 16 inches long, 3 to 9 inches broad; spines over the entire joint or restricted to the upper one-third or sometimes one-half or two-thirds (rarely none), reddish brown or dark brown, in some forms grayish or sometimes lighter red or red-and-gray, 1 to 6 or 10 per areole, spreading at right angles or some of them deflexed, straight or bent downward or sometimes curved or twisted, the longer ones usually 1 or 2 to 3 inches long, the larger ones basally 1/32 to 1/16 inch broad, some clearly flattened, narrowly elliptic or broader in cross section, not barbed; flower 2½ to 3¼ inches in diameter, 2½ to 3 inches long; petaloid perianth parts yellow or the bases red, obovate or cuneate-obovate, 1 to 1½ inches long, ¾ to 1 inch broad, truncate or rounded, minutely somewhat denticulate; fruit wine-color or purplish, fleshy at maturity, smooth, the areoles not prominent, usually obovate, sometimes elongate, 1¼ to 2½ inches long, ¾ to 1½ inches in diameter, the apical cup not deep, persisting until winter; seeds light tan or grayish, irregularly discoid, about 3/16 inch long.

Opuntia phaeacantha is highly variable, and there are several geographical varieties. Many segregate species have been proposed, but these have been based upon individual variation or upon individuals obtained from hybrid swarms resulting from inter-

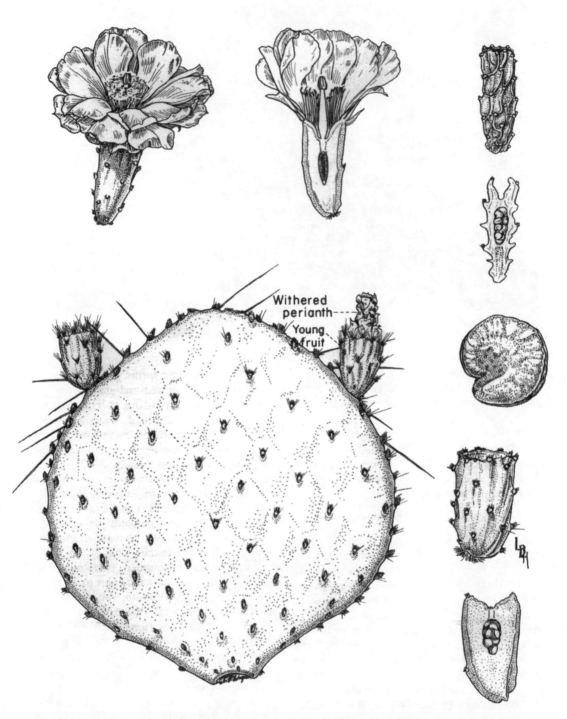

Withered
perianth
Young
fruit

Fig. 1.69. Purple prickly pear, *Opuntia violacea* var. *santa-rita. Above,* flower in whole view
and in longitudinal section. *Right margin,* fruits in side view and longitudinal section, the upper
ones having dried long after maturity; seed. *Below,* joint with young fruits.

breeding with other species or between the varieties. Hybridizing is common between all the geographically contiguous varieties and where this species occurs with the following as well as other species:

Opuntia macrorhiza on the high plains of northern Arizona and New Mexico (with var. *phaeacantha*) and on the Great Plains (with vars. *camanchica* and *major*).

Opuntia littoralis var. *Martiniana* in northwestern Arizona and other varieties in California. Hybridizing is with var. *major* or var. *discata*.

The varieties described for this species are by no means discrete, but they represent clear tendencies associated with geographical areas.

A winter characteristic is the appearance of a bright red border on the margin or of red streaks on the face of the joint. During the growing season the plant is green.

25A. Var. *phaeacantha*
(Table 9)

DISTRIBUTION: Sandy or rocky soils of hills, valleys, and canyons in woodlands and grasslands at 4,500 to 6,000 or 8,000 feet elevation. Mostly in the Juniper-Pinyon Woodland but in the Great Plains Grassland and the lower edge of the Rocky Mountain Montane Forest. Arizona along and north of the Mogollon Rim and along the eastern boundary in Cochise County (rare). Utah in the La Sal Mountains; Colorado from the vicinity of Grand Junction and along the southern boundary eastward to the edge of

TABLE 9. CHARACTERS OF THE VARIETIES OF OPUNTIA PHAEACANTHA

	A. Var. **phaeacantha**	B. Var. **laevis**	C. Var. **major**	D. Var. **discata**
Joint shape	Obovate.	Narrowly obovate.	Broadly obovate or nearly orbiculate.	Orbiculate to elliptic.
Joint size	4 to 6 inches long, 3 to 4 inches broad.	6 to 10 inches long, 4½ to 6 inches broad.	Usually 5 to 10 inches long, 4 to 8 inches broad.	Mostly 9 to 12 or 16 inches long, 7 to 9 inches broad.
Distribution of spines on joint	Usually over the upper three-fourths or more.	Spines **none**, except in intergrades to other varieties.	Over usually the upper one-half, one-third, or less of the joint.	In all or all but a few basal areoles on the joint.
Spine color	Dark or sometimes light brown.	. .	Dark brown.	White or ashy gray.
Maximum spine length	1½ or 1¾ to 2½ inches.	. .	1¼ to 2¼ or 2¾ inches.	1 to 2 or rarely 2½ or even 3 inches.
Maximum spine width	1/32 to 1/24 inch.	. .	Usually 1/24 to 1/16 inch.	1/24 to usually 1/16 inch.
Spines per areole	Above 3-5 or 9, below 1-2.	. .	Usually 1-3.	1-4 or rarely 10.
Fruit type	Obovoid, 1¾ to 2¾ inches long.	Cylindroidal, the base narrowed, 2¾ to 3¼ inches long.	As in var. **phaeacantha** but a little larger.	As in var. **phaeacantha** but a little larger.
Vegetation type	Mostly Juniper-Pinyon Woodland.	Arizona Desert.	Mojavean, Arizona, and Chihuahuan deserts and Desert Grassland.	Mojavean, Colorado, Arizona, and Chihuahuan deserts; Desert Grassland, and other grass lands.
Geographical distribution	Northern and eastern Arizona at higher altitudes. Southeastern Utah; southern Colorado; mostly in the northwestern quarter of New Mexico but occasional southward; Texas in the Big Bend region.	Canyons and cliffs. Arizona in Gila and Pima counties.	Throughout Arizona, especially at lower elevations. The Californian deserts to southwestern Utah, southern Kansas and central Texas.	Arizona mostly at lower elevations. California in the mountains bordering the deserts and eastward to Utah (Washington County) and to Texas (eastern and southern hills of the Edwards Plateau). Mexico in Sonora and Chihuahua.

Fig. 1.70. *Opuntia phaeacantha*. The typical variety occurs in the high plateau country of northern Arizona, northwestern New Mexico, and adjacent Utah and Colorado. Note the relatively small, narrow joints and the long, brown spines over most of the joint. (Photograph by David Griffiths.)

SYNONYMY: Opuntia phaeacantha Engelm. Type locality: Santa Fe, New Mexico. *Opuntia dulcis* Engelm. *Opuntia Lindheimeri* Engelm. var. *dulcis* Coulter. *Opuntia Engelmannii* Salm-Dyck var. *dulcis* Engelm. Type locality: near "El Paso?, Texas?" *Opuntia zuniensis* Griffiths. Type locality: near Zuñi, New Mexico.

25B. Var. **laevis**
  (Coulter) L. Benson, **comb. nov.** (p. 21)
  (Table 9)

DISTRIBUTION: Ledges of cliffs and steep canyon walls or sometimes in protected places in good soil under trees in canyon bottoms in the deserts, grasslands, and woodlands at 2,500 to 3,000 feet elevation. Upper part of the Arizona Desert and the Desert Grassland and lower edge of the Southwestern Oak Woodland. Arizona from Pinal County near Sacaton and from Gila County to Pima County.

Spineless cacti are destroyed by cattle, and survival is dependent upon ability to grow on inaccessible ledges or in cracks between rocks or where there is protection by trees or brush. Rodents also eat the plants, especially during prolonged dry periods, when they depend upon them for water.

SYNONYMY: Opuntia laevis Coulter. Type locality: Arizona.

Name Applied to a Hybrid

Opuntia cañada Griffiths. *Opuntia laevis* Coulter var. *cañada* Peebles. Type locality: Santa Rita Mountains, Arizona. (Var *laevis* x *major*).

Opuntia cañada Griffiths is intermediate between *Opuntia phaeacantha* var. *major* and var. *laevis*. The fruits are of the obovoid type of the former variety, and the spines are similar to the latter but distributed over the entire joint. The joints are green as in var. *laevis*. This form has been found in the Santa Rita, Santa Catalina, and Baboquivari mountains. According to Professor Thornber, the type collection was made at the dam just above the Forest Service experiment station in Florida (Stone Cabin) Canyon in the Santa Rita Mountains, although Griffiths mentioned only "foothills of the Santa Rita Mountains." Specimens from the Florida Canyon dam include typical var. *laevis* (*L. Benson 9883*) and the *Opuntia cañada* type (*L. Benson 9884, Darrow in 1939*). (Specimens, *Pom.*)

Opuntia gilvescens Griffiths. Joints obovate, usually spiny over half or two thirds of the surface; base of the fruit not markedly constricted. Santa Rita Mountains. This plant is almost identical with

the Great Plains; northern New Mexico in the Rocky Mountain region and at mostly higher altitudes southward; Texas (rare) in Jeff Davis, Brewster and Terrell counties. Characters of this variety occur individually in some specimens from southern Arizona.

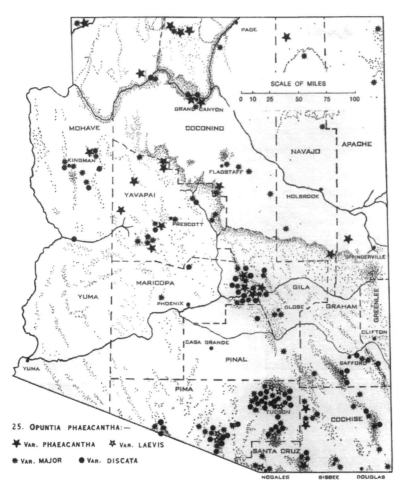

Fig. 1.71. The documented distribution of *Opuntia phaeacantha*, according to its varieties.

25. OPUNTIA PHAEACANTHA:—

✿ VAR. PHAEACANTHA   ✿ VAR. LAEVIS

✴ VAR. MAJOR   ● VAR. DISCATA

Opuntia cañada Griffiths, first collected in the same vicinity. The lack of spines on the lower (basal) surface of the joint is the only difference worthy of note, and even this is variable in the mixed population in Florida Canyon. One sheet of *L. Benson 9884* is this plant.

25C. Var. *major*
Engelm.
(Table 9)

DISTRIBUTION: Rocky, gravelly, or sandy soils of hills, valleys, or flats mostly in the Arizona and Chihuahuan deserts and the Desert Grassland and also in the lower portion of the Juniper-Pinyon Woodland or the Mojavean Desert or the desert-edge phase of the California Chaparral or the lower edge of the Sierran Montane Forest and sparingly in other vegetation types. Arizona above (uncommon) and below (common) the Mogollon Rim from Mohave County to the Baboquivari Mountains in Pima County and eastward and also on the Colorado River at Yuma. California from San Luis Obispo County eastward through Cuyama Valley to the desert side of the mountain axis from Los Angeles County to San Diego County, and in the eastern Mojave Desert (rare); southern Nevada, southern margin of Utah; southern Colorado (occasional); New Mexico, chiefly in the southern half of the State; South Dakota; western Kansas (rare); western Oklahoma (rare); Texas Panhandle and from El Paso to the Edwards Plateau and rare in the hill country eastward and southeastward. Mexico in Sonora and Chihuahua.

Fig. 1.72. *Opuntia phaeacantha* var. *laevis*. A spineless variety occurring in southern Arizona, usually restricted to rocky ledges in the mountains. (Photographs by H. L. Shantz.)

In California the plant has been overlooked or confused with other species. In some books it appears under a synonym, *Opuntia mojavensis,* and mountain forms have been confused with *Opuntia littoralis* var. *Piercei,* with which it intergrades.

SYNONYMY: Opuntia phaeacantha Engelm. var. *major* Engelm. Type locality: Santa Fe, New Mexico. *Opuntia Engelmannii* Salm-Dyck var. *cyclodes* Engelm. & Bigelow. *Opuntia Lindheimeri* Engelm. var. *cyclodes* Coulter. *Opuntia cyclodes* Rose. Type locality: On the upper Pecos River, New Mexico. *Opuntia mojavensis* Engelm. *Opuntia phaeacantha* Engelm. var. *mojavensis* Fosberg. Type collection: Mojave River, California. *Opuntia phaeacantha* Engelm. var. *brunnea* Engelm. Type locality: El Paso, Texas. *Opuntia arizonica* Griffiths. Type locality: near Kirkland, Arizona. *Opuntia gilvescens* Griffiths. Type locality: Santa Rita Mountains, Arizona. *Opuntia Toumeyi* Rose. Type locality: Desert Laboratory, Tucson, Arizona. *Opuntia Blakeana* Rose. Type locality: Desert Laboratory, Tucson, Arizona. *Opuntia Gregoriana* Griffiths. Type locality: El Paso, Texas. *Opuntia confusa* Griffiths. Type locality: Tumamoc Hill (Desert Laboratory), Tucson, Arizona. *Opuntia expansa* Griffiths. Type locality: Anton Chico, New Mexico. *Opuntia curvospina* Griffiths. Type locality: between Nipton, California, and Searchlight, Nevada. *Opuntia flavescens* Peebles. *Opuntia Engelmannii* Salm-Dyck var. *flavescens* L. Benson. Type locality: near Sells, Pima County, Arizona.

25D. Var. *discata*
(Griffiths) Benson & Walkington
Engelmann prickly pear (Table 9)

DISTRIBUTION: Sandy soils of plains, washes, hills, valleys, and canyons in the desert and grasslands at 1,500 or commonly 2,000 to 4,000 or 5,000 feet elevation. Mostly in the Arizona Desert and the Desert Grassland but also the Mojavean Desert, the upper edge of the Colorado Desert, and the Chihuahuan Desert and other vegetation types. Arizona in the Mojavean and especially the Arizona deserts and the Desert Grassland, occasional from Mohave County eastward through northern Arizona and common southeastward just below the Mogollon Rim to Greenlee County and southward to the Baboquivari Mountains in Pima County and to Cochise County. California mostly on the desert side of the San Bernardino, San Jacinto, and Laguna mountains and in the eastern Mojave Desert; reported from Nevada in Clark County; Utah in Washington County

Fig. 1.73. *Opuntia phaeacantha* var. *major.* Note the restriction of the spreading, moderately long, brown spines largely to the upper portion of the joint, those on the flat sides of the joints (when present) tending to be shorter, fewer, and lighter in color. (Photograph by David Griffiths.)

This is the most abundant prickly pear in the deserts of the Southwest. On the Colorado Plateau and in the southern Rocky Mountains it is replaced largely by var. *phaeacantha,* a more localized variety of the highlands. Var. *phaeacantha* is most common in northwestern New Mexico.

Fig. 1.74. Engelmann prickly pear, *Opuntia phaeacantha* var. *discata* (long known mistakenly as *Opuntia Engelmannii*).

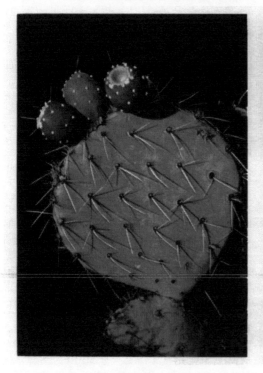 

Fig. 1.75. Engelmann prickly pear, *Opuntia phaea-cantha* var. *discata*. Note the occurrence of relatively short, white spines over the entire joint. (Photograph by David Griffiths.)

(rare); New Mexico below 5,000 feet from Grant and Hidalgo counties to Lincoln and Otero counties and in Bernalillo County; Texas west of the Pecos River and sparingly eastward to the hills east of the Edwards Plateau. Mexico in Sonora, Chihuahua, and Coahuila.

This is the largest and, in especially southern Arizona, one of the best-known native prickly pears of the Southwestern Deserts of the United States. It is variable in habit of growth, shape and

size of joints, and size and distribution of spines. It is almost always found growing with var. *major,* which has longer brown spines restricted largely to the upper part of the narrower joint. Almost everywhere there are numerous intergrading forms with many character recombinations. Var. *discata* is rarely stable but apparently a fringe-population extreme tied in closely with the more abundant and wide-ranging var. *major.*

Specific rank is untenable, and the widely used name *Opuntia Engelmannii* was not applied to this taxon but to the cultivated spiny mission cactus, a form of *Opuntia Ficus-Indica* known as "Opuntia megacantha," from Mexico. Unfortunately, the long-established designation is upset upon the basis of both classification and nomenclature.

SYNONYMY: Opuntia procumbens Engelm. & Bigelow. Type locality: Aztec Pass, western Yavapai County, Arizona. *Opuntia angustata* Engelm. & Bigelow. *Opuntia phaeacantha* Engelm. var. *angustata* Engelm. ex W. T. Marshall, incorrectly ascribed to Engelmann. Type locality: Bill Williams River, Arizona. *Opuntia discata* Griffiths. *Opuntia Engelmannii* Salm-Dyck var. *discata* C. Z. Nelson. *Opuntia phaeacantha* Engelm. var. *discata* Benson & Walkington. Type locality: Santa Rita Mountains, Pima County, Arizona. *Opuntia megacarpa* Griffiths. *Opuntia Engelmannii* Salm-Dyck var. *megacarpa* Fosberg. *Opuntia occidentalis* Engelm. & Bigelow var. *megacarpa* Munz. Type locality: near Banning, California. *Opuntia magnarenensis* Griffiths. Type locality: southeast of Kingman, Arizona. *Opuntia xerocarpa* Griffiths. Type locality: southeast of Kingman, Arizona. *Opuntia Woodsii* Backeberg. Type locality "USA (Nevada)."

26. *Opuntia chlorotica*
     Engelm. & Bigelow
     Pancake pear

Arborescent or an erect shrub 2 or 3 to 6 feet high, 3 to 4 feet in diameter; trunk about 1 foot long, 3 to 8 inches in diameter; joints blue-green, orbiculate to broadly obovate, 6 to 8 inches long, 5 to 7 inches broad; spines in all but the basal areoles of the joint, light yellow to sometimes straw color but blackened or dirty gray in age or in old dried specimens, 1 to 6 per areole, all turned downward, straight or curving at the extreme bases, 1 to 1⅝ inches long, basally about 1/24 or sometimes 1/48 inch broad, flattened, tapering their entire length, narrowly elliptic to linear in cross section, not markedly barbed; flower about 1½ to 2½ inches in diameter; petaloid perianth parts light yellow, with a reddish flush, broadly obovate to cuneate-obovate,

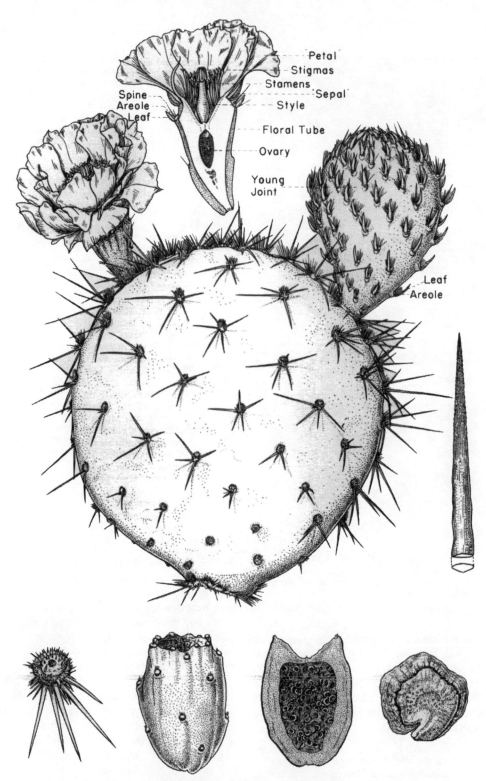

Fig. 1.76. Engelmann prickly pear, *Opuntia phaeacantha* var. *discata*. *Above,*
a young joint with leaves and an old one without them; a flower and a longitudinal
section of a flower. *Right,* a spine enlarged. *Below,* a spine cluster and an areole;
fruit in side view and longitudinal section; seed.

Fig. 1.77. Pancake pear, *Opuntia chlorotica*. (Photographs by David Griffiths.)

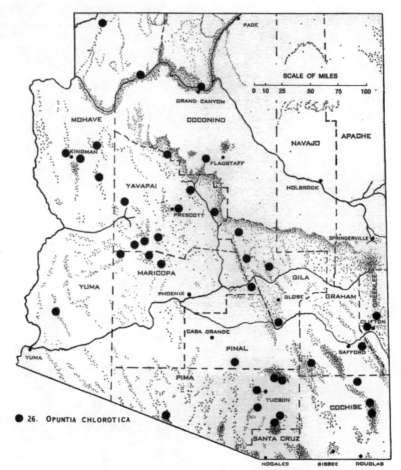

Fig. 1.78. The documented distribution of *Opuntia chlorotica*.

● 26. OPUNTIA CHLOROTICA

¾ to 1¼ inches long, ½ to ¾ inch broad, truncate or rounded and mucronate, essentially entire; fruit grayish tinged with purple, fleshy at maturity, subglobose to ellipsoid, with no spines, 1½ or up to 2½ inches long, ¾ to 1⅜ inches in diameter, the apical cup bowl-like or sometimes shallow, maturing the summer or fall after flowering; seeds light tan, nearly elliptic but asymmetrical, smooth, 3/32 to 1/8 inch long.

DISTRIBUTION: Sandy or rocky soils of ledges, slopes, canyons, or flats in the desert or just above it at 2,000 or mostly 3,000 to 4,000 or rarely 6,000 feet elevation. Colorado Desert (rare); Mojavean Desert; Arizona Desert; Southwestern Oak Woodland and Chaparral; Desert Grassland. Arizona from Mohave County to the lower parts of northern Coconino County and southeastward be-low the Mogollon Rim to Graham, Pima, and Cochise counties. California in the southern and eastern Mojave Desert and the western edge of the Colorado Desert and just above it; southern Nevada in the Charleston Mountains, Clark County; southwestern New Mexico. Mexico in northern Baja California and Sonora.

Plants intermediate between this variety and *Opuntia violacea* or possibly var. *santa-rita* occur in the Perilla, Swisshelm, and Huachuca mountains and east of the Mule Mountains in southeastern Arizona.

Opuntia chlorotica is usually a mountain plant, but it occurs near Aguila on low flats covered with tobosa grass (*Hilaria mutica*). Some specimens in the vicinity are more than 6 feet high and with more than 200 joints.

SYNONYMY: Opuntia chlorotica Engelm. & Bigelow. Type locality: Mojave River, California.

2. CEREUS

Stems elongate, branching slightly to freely, cylindroid to prismatic, at maturity 15 to 100 times as long as their diameter, 1 to 50 feet long, ¼ inch to 2½ feet in diameter; ribs 3 to more than 20; tubercles coalescent into uniform ribs. Leaves not discernible in the mature plant. Spines smooth, gray, yellow, straw-color, tan, brown, red, or white; central spines not usually differentiated, the spines 1 to many per areole, straight, usually 1/16 to 3 inches long, basally usually 1/240 to 1/16 inch in diameter, needlelike, broadly elliptic in cross section. Flowers and fruits produced on the old growth of preceding years and therefore below the growing apex of the stem or branch, developing in a felted area within at least the edge of the spine-bearing part of the areole or merging into it. Flower usually 1 to 6 or 9 inches in diameter; floral tube above the ovary ranging from almost obsolete to funnelform or long and tubular. Fruit fleshy at maturity, usually pulpy, often edible, with or without tubercles, scales, spines, hairs, or glochids, usually orbicular to ovoid or ellipsoid, usually ½ to 3 inches long, indehiscent, the floral tube deciduous or persistent. Seeds usually black, smooth to reticulate or papillate, longer than broad (length being hilum to opposite side), usually 1/24 to 1/12 inch long; hilum obviously basal, sometimes oblique.

DISTRIBUTION: An undetermined large number of species from California, Arizona, New Mexico, Texas, and Florida to South America. Sixteen native or introduced in the United States. Five in Arizona.

KEY TO THE SPECIES

1. None of the roots forming tubers; mature stems ¾ inch to 2 feet in diameter; spines or some of them *nearly always* ½ inch long or longer.
    2. Tree with a trunk bearing usually only 1 to 5 or 15 branches, but gigantic plants 25 to 50 feet high having a larger number; stems 6 inches to 2½ feet in diameter, narrower at ground-level than above; superior floral tube slender, at least 1½ inches long; ovary completely covered at flowering time by green scales, these *not* enlarging with growth of the fruit but the fruit bearing obvious scales, the areoles in their axils *not* bearing discernible spines, bristles, or hairs; fruit splitting lengthwise along 3 lines, green outside and red inside.   1. *Cereus giganteus,* page 108

2. Plant (in the United States) *not* treelike, the axis branching many times at ground-level and above; stems ⅜ to 2 inches in diameter, at least as large at ground-level as above; superior floral tube below the attachment of the stems relatively broad, *not* more than ⅝ inch long; ovary *not* completely covered at flowering time by green scales, the areoles bearing spines, bristles, or hairs, the fruit *not* scaly; fruit opening irregularly, if at all.
    3. Flower solitary, that is, only 1 per areole; areoles of the lower and upper parts of the stem producing similar spines, ½ to 1 inch long, not bulbous-based; fruit with a dense but deciduous cover of spines.
        2. *Cereus Thurberi,* page 111
    3. Flowers 2 to several in an areole (or rarely only 1 in some areoles); areoles of the lower part of the stem (in the species occurring in Arizona) spineless or producing about 15 stout spines approximately ½ inch long and with conspicuous bulbous bases, each areole of the upper part of the stem producing about 30 slender spines ½ to 1½ inches long and without bulbous bases; fruit with scales but spineless or rarely with a few basal spines.
        3. *Cereus Schottii,* page 114
1. Either the taproot or a cluster of principal roots forming one or more large tubers, these fleshy and outwardly resembling sweet potatoes and internally raw white potatoes; stems slender, basally about 3/16 to ¼ inch in diameter, enlarging upward to ¼ to ½ inch, with as many as 8 branches above; spines (Arizona species) slender and weak, sometimes with bulbous bases, not more than ⅜ inch long.
    2′. Stems 4-6-ribbed, the ribs conspicuous, the intervening troughs deep; the style and the stamens not projecting above the perianth parts; tuber one, this a great enlargement of the taproot weighing from 1½ to 43 pounds; spines mostly parallel to the stem-surface, but one or more nearly at right angles; superior floral tube 4 to 6 inches long.   4. *Cereus Greggii,* page 117
    2′. Stems 6-10-grooved, the ridges only about 1/24 to 1/16 inch high; the style and the stamens at least often projecting beyond the perianth parts; tubers several, one developed from each of several major roots; spines all lying parallel to the stem-surface; superior floral tube 2 to 3 inches long.
        5. *Cereus striatus,* page 119

LIST OF SPECIES AND VARIETIES

The genus has been subdivided into proposed separate "genera" according to various philosophies and mostly artificial approaches, that is, classification upon predetermined grounds. An arbitrary value has been assigned to some characters; others have been ignored. Thus the plants have been fit to the theory instead of the reverse.

Cereus doubtless needs to be subdivided, but this will become feasible only after laborious and very extensive gathering of data. The presently available information is inadequate, and the way forward is not clear. Engelmann noted that he would be able perhaps to classify *Cereus* in the next world. The writer may be in the same position.

1. *Cereus giganteus*
   Engelm.
   Saguaro, giant cactus

Tree 9 to 50 feet high, the trunk 1 to 2½ feet in diameter, for many years unbranched, but later with 1 to 5 or up to about 20 branches which bend or curve abruptly upward, produced well above the base of the stem, with little rebranching, green; ribs about 12 to 30, prominent, mostly about 1 to 1½ inches high; spines rather dense, gray or with a pinkish tinge, 15 to 30 per areole, on young stems and the lower portions of older trunks the longer ones deflexed, straight, up to 3 inches long, on the upper parts of the older stems spreading and 1 to 1½ inches long, basally up to 1/20 inch in diameter, needle-like or on young stems flattened or angular; flower nocturnal but remaining open the next day, about 2 inches in diameter; petaloid perianth parts white, the largest ones more or less obovate, up to about 1 inch long, about ⅝ inch broad, rounded, undulate; fruit green tinged with red, fleshy at maturity, with scales (those covering the floral tube at flowering time not having grown as the fruit developed) but otherwise smooth, obovoid to ellipsoid, 2 to 3 inches long, about 1 to 1¾ inches in diameter, with the apical cup not prominent, maturing in July, dehiscent along usually three regular vertical lines and exposing the conspicuous red lining; seeds black, about ½ inch long.

DISTRIBUTION: Rocky or gravelly soils of foothills, canyons, and benches and along washes in the desert at 600 to 3,600 feet elevation. Arizona Desert and the upper edge of the Colorado Desert. Arizona from southern Mohave County to Graham County (Aravaipa), Yuma County, and Pima County. California near the Colorado River from the Whipple Mountains to the Laguna Dam (a few individuals). Mexico in Sonora.

According to letters from Dr. Forrest Shreve, Carnegie Institution of Washington Desert Laboratory, Tucson, to Dr. F. V. Coville, United States National Herbarium, a great saguaro near Sabino Canyon, northeast of Tucson, fell in 1915. This plant weighed 9 tons, the estimate having been based upon careful measurements to determine volume in relationship to the weight of samples. The letters are accompanied by photographs of the plant taken by Dr. D. T. McDougal in 1903 and 1911.

The name is pronounced sa-war-o, and it has been transliterated in various ways. The English name, giant cactus, is also in common usage, particularly in the United States outside of Arizona, where the name saguaro is all but unknown.

The flowers are white, and the "red flowers" reported annually in midsummer are the opened fruits, which have red pulp. This pulp is an important article of diet for many species of birds. The ranges of the white-wing dove and the elf owl (no larger than a sparrow) are almost identical with the distributional area of the saguaro. However, the elf owl is concerned with cover rather than food, for it nests in abandoned woodpecker holes in the saguaro trunks.

Indians gather the fruit for making conserves and beverages, and for the seeds, which are rich in fats and which at one time were used extensively for chicken feed. The woody carcass of the plant is useful to man in construction of shelters, light corrals, and novelty furniture and trinkets. The heavy wood is so strongly impregnated with mineral crystals that it dulls cutting tools quickly.

The saguaro is a characteristic plant of Arizona and Sonora and is one of the most striking features of the desert landscape. In the eastern and northern parts of the distributional range it is found on the south-facing slopes, where it is best protected from the winter cold. In the forests south of Ajo, the plants are as abundant on the north slopes as on the south slopes. In warmer areas in Mexico they grow only on north slopes, as do many other Northern Hemisphere species at the southern limits of their ranges.

On the western edge of the distributional area the altitude is less and temperature is not a critical factor. Because the rainfall is less than 5 inches, the saguaros are restricted to drainage channels. The distribution of the species is particularly striking west of O'Neill Hills where long rows of saguaros follow threadlike washes many miles in length, and only barren desert and creosote bush areas lie between the washes.

Anchorage is an important factor in the habitat of the plant. In the range of the saguaro strong

Fig. 2.1. Saguaro or giant cactus, *Cereus giganteus*. (Photograph by Ray Manley.)

Fig. 2.2. Forest of saguaros, *Cereus giganteus*. (Photograph by R. B. Streets.)

Fig. 2.3. Flowers of the saguaro, *Cereus giganteus*. The flowers open in the late evening, remaining open through the night and usually (according to temperature) much of the following day. See also Fig. 0.4. (Photograph by Robert H. Peebles.)

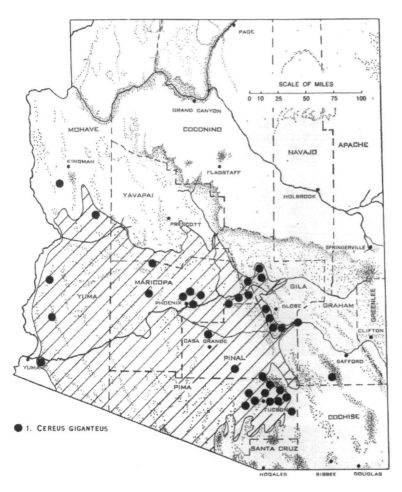

Fig. 2.4. The documented distribution of *Cereus giganteus*. Additional distribution based upon field observation is indicated.

winds are common in rainy weather, when the ground is soft, and sometimes the root anchorage is not adequate for supporting several tons of plant material swaying in the wind. In one cactus forest in Sonora fully half the giant cacti had been blown over by the wind and were half buried in alluvial soil. All were pointing in the same direction.

Extensive forests of saguaros are common in southern Arizona. One of the most interesting is reached by traveling eastward on Speedway or Broadway in Tucson for about 14 miles. This area has been set aside as a recreation area, and it is known as the Saguaro National Monument. Various species of cacti besides the saguaro are abundant there. Other excellent forests are at the south end of Santa Rosa Valley in western Pima County and in the Tucson Mountains.

SYNONYMY: Cereus gigantens [*giganteus*] Engelm. *Pilocereus Engelmannii* Lemaire, *nom. nov. Pilocereus giganteus* Rümpler. *Carnegiea gigantea* Britton & Rose. Type locality: Coolidge Dam, Arizona.

2. *Cereus Thurberi*
Engelm.
Organ-pipe cactus, pitahaya

Large plant with numerous columnar branches, in the United States these arising from ground-level and resembling the pipes of an organ, commonly 9 to 20 feet high, 6 to 18 feet in diameter; trunk none or (in Mexico) sometimes present, short; joints green, cylindroidal, usually (unless the plant has been killed back by frost) 9 to 20 feet long, 4 to 6 or 8 inches in diameter; ribs 12 to 19, $\frac{3}{8}$ to $\frac{1}{2}$ inch high; spines rather dense on the stem, dirty-gray, black, or brownish, 11 to 19 per areole, spreading in all directions, straight, up to 1/2 inch long, basally up to 1/48 inch in diameter, needlelike, nearly circular in cross section; flower nocturnal (or

Fig. 2.5. *Cereus. Above,* fruits on a branch of the saguaro, *Cereus giganteus. Below,* desert nightblooming cereus, *Cereus Greggii* var. *transmontanus,* with 31 open flowers. (Photograph, *below,* by Robert A. Darrow.)

Fig. 2.6. Organ pipe cactus, *Cereus Thurberi.* (Photograph by Robert H. Peebles.)

Fig. 2.7. Organ pipe cactus, *Cereus Thurberi.* Branch with flowers and a young and an old fruit. (Photograph by Robert H. Peebles. From Thomas H. Kearney and Robert H. Peebles, *Arizona Flora.* University of California Press, Berkeley. 1951, 1960; used with permission.)

remaining open the next day), about 2½ to 3 inches in diameter; floral tube with short scale-leaves overlapping like shingles; petaloid perianth parts lavender, the margins white, the largest ones oblanceolate, ¾ to 1 inch long, ⅜ to ½ inch broad, spreading, the tips recurved, acute, entire; fruit red, fleshy at maturity, edible, with a dense but deciduous cover of spines, spheroidal, 1¼ to 3 inches in diameter, with the apical cup not prominent, maturing in the late summer or fall; seeds black, 1/12 to 1/10 inch long.

DISTRIBUTION: Rocky or sandy hills, mesas, and valleys in the desert at 1,000 to 3,500 feet elevation. Arizona Desert and (in Mexico) the other Sonoran Deserts. Arizona in Pinal and Pima counties from the Growler Mountains southwest of Ajo to the Picacho and Roskruge mountains. Mexico in Baja California and western Sonora.

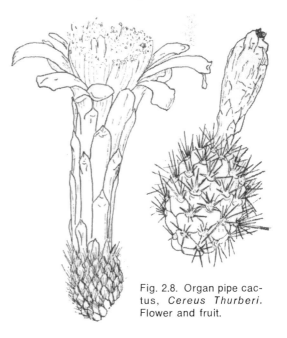

Fig. 2.8. Organ pipe cactus, *Cereus Thurberi.* Flower and fruit.

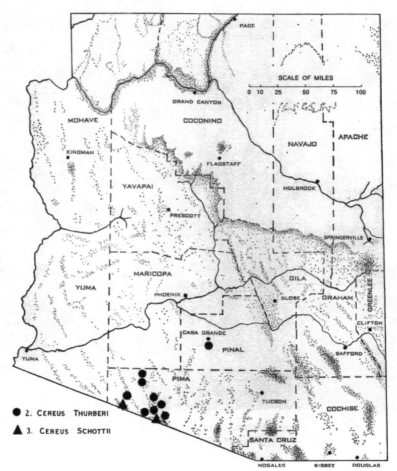

Fig. 2.9. The documented distribution of *Cereus Thurberi* and of *Cereus Schottii*.

The fruit is palatable, and it is gathered in quantity by the Papago Indians, who move into the Pitahaya areas for the harvest.

Arizona plants are more branched than those of the same age occurring in Sonora where the climate is milder. The organ pipe cactus is sensitive to frost, and, although the plant seldom is killed, the tender growing tips may be. After killing of the tips, new shoots are sent up from near the ground level. As in other cacti of *Cereus* and related genera, once the stem tip is killed, its elongation ceases. However, the stem may continue to live for an indefinite length of time, and it may increase remarkably in girth.

SYNONYMY: Cereus Thurberi Engelm. *Pilocereus Thurberi* Rümpler. *Lemaireocereus Thurberi* Britton & Rose. *Neolemaireocereus Thurberi* Backeberg. *Marshallocereus Thurberi* Backeberg. Type locality: Bacuachi, a small town on the road to Arispe, Sonora.

3. *Cereus Schottii*
Engelm.
Senita

Plant with elongated ascending branches, 6 to 21 feet high and 6 to 15 feet in diameter; trunk none; joints green, several feet long, about 4½ to 5 inches in diameter; ribs 5 to 9, mostly 6 to 7, prominent, about 1 inch high; spines of the upper part of the stem gray or with a pinkish tinge, about 30 to 50 per areole, tending to be turned downward, bristlelike, twisted, up to 3 inches long, strongly flattened, striate, those of the lower part of the stem gray, about 8 or 10 per areole, 1 central and 7 to 9 radial, straight, short, bulbous-based, about ⅜ inch long, stout, about 1/16 inch in diameter just above the bulbous bases, circular to angular in cross

Fig. 2.10. Senita, *Cereus Schottii*. (Photographs by H. L. Shantz.)

Fig. 2.11. Desert night-blooming cereus, *Cereus Greggii* var. *transmontanus*. (Photographs by H. L. Shantz.)

section, 1 to several per areole; flowers nocturnal, with an unpleasant odor, 1 to 1½ inches in diameter; petaloid perianth parts pink, the largest ones oblanceolate, up to 1 inch long, up to ¼ inch broad, mucronate, sparsely denticulate; fruit red, fleshy at maturity, with a few scales, nearly globular to ovoid, about 1 to 1¼ inches long, ¾ to 1 inch in diameter, with the apical cup not prominent, ripening in the fall, bursting irregularly; seeds 1/10 to 1/8 inch long.

DISTRIBUTION: Heavy or mostly sandy soils of valleys and plains in the desert at 1,500 feet or less elevation. Sonoran Deserts. Middle southernmost Arizona at the extreme southern edge of the Organ Pipe Cactus National Monument, Pima County. Mexico in Baja California and Sonora.

Not more than 50 individual plants are now native in the United States. Probably the number never exceeded 100.

Specimens on the grassy plains of Sonora have several hundred branches, but the largest counted in the United States (Abra Valley, southeast of Ajo) had 134. In age a single clump may include several independent plants, but all of these have originated vegetatively usually from a single plant.

The slender spines of the upper portion of each plant are conspicuous, and they resemble gray hair.

SYNONYMY: Cereus Schottii Engelm. *Pilocereus Schottii* Lemaire. *Lophocereus Schottii* Britton & Rose. Type locality: "Sonora, toward Santa Magdalena."

4. *Cereus Greggii*
Engelm.
Desert night-blooming cereus

Slender and exceedingly inconspicuous plants with the appearance of dead sticks or of branches of the creosote bush; stems erect or sprawling, 1 to 2 or 8 feet * long, up to ½ inch in diameter, very slender below and only about ¼ inch in diameter; stems not branched or with up to five or more branches, strongly 4-, 5-, or 6-ribbed above, minutely and densely canescent; root very large and turnip-like; spines abundant but small, about 11 to 13 per areole (about 6 or 8 upper ones only 1/32 inch long, dark; about 3 to 5 lower ones about ⅛ inch long, 2 to 4 of them white, the others dark), lying mostly parallel to the surface, straight, basally bulbous, acicular, nearly circular in cross section; flower nocturnal, 2 to 3 inches in diameter, 6 to 8½ inches long; floral tube 4 to 6 inches long, with minute spines but no scale-leaves; petaloid perianth parts white, the largest ones narrowly oblanceolate, up to 2 inches long, up to ½ inch broad, mucronate, entire; fruit bright red at first, dull in age, fleshy at maturity, with the areoles bearing short spines, ellipsoid, 2 to 3 inches long, 1 to 1½ inches in diameter, without an apical cup, ripening in late summer; seeds ⅛ inch long.

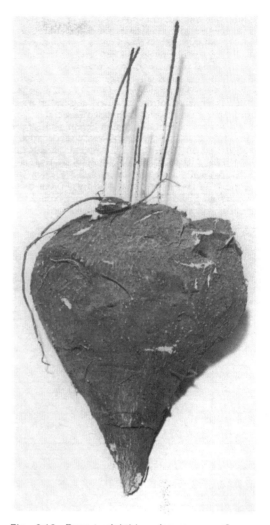

Fig. 2.12. Desert nightblooming cereus, *Cereus Greggii* var. *transmontanus*. The proportionately enormous tuber, weighing up to 43 pounds; with young stems growing from the top. (Photograph by Robert H. Peebles.)

Birds open the fleshy fruit from the side of the wall, and few, if any, of the seeds fall naturally to the ground. Since nearly all the seeds are eaten by birds, the distribution of the cactus in the desert is determined chiefly by the movements and roosting habits of the species that feed on the fruits. Consequently the night-blooming cereus occurs chiefly under trees and shrubs in the alluvial bottomlands or on the hillsides. The plant is dependent

upon the surrounding woody vegetation for protection of the stem from animals, for the lower part of it is brittle and readily broken.

Flowering occurs primarily on one or two nights late in May or in June. However, a few flowers may appear earlier or later, and the total flowering period occupies nearly one month. Flowers open just after dusk when the petals and other floral structures begin a series of spasmodic jerks. The perfume of the blossom is liberated in profusion, and a single flower may scent the air for 100 feet. Normally the blooming period is over by about seven o'clock the following morning, or later if the

---

* A record is eleven feet ten inches; cultivated plant, *Mrs. William H. Kitt, US* (photo).

weather is cool, but the flower may be removed from the plant and kept open for some time if it is placed in the dark. Sometimes it may be preserved for a while in water in a refrigerator. Usually a single plant produces only a few flowers each season, but one unusually large plant in Tucson produced forty-four flowers in 1939 and thirty-one in 1940. Another plant on the desert at Sacaton was photographed with twenty-four open flowers.

The root of the night-blooming cereus is a large tuber shaped like a turnip. Twenty-seven roots weighed at the University of Arizona varied in weight from 1½ to 43 pounds. Indians and Mexicans occasionally slice the tubers and fry them in deep fat. The taste is like that of a turnip, but it is less pungent. Slices of the tubers also are bound on the chest as a supposed cure for congestions.

4A. Var. *transmontanus*
    Engelm.

Areoles circular or nearly so, about 1/16 inch in diameter; flower up to about 3 inches in diameter, up to 8½ inches long; bristles from the upper areoles of the floral tube ⅗ to 1¼ inches long; petaloid perianth parts up to 3 inches long, relatively narrow, narrowly lanceolate.

DISTRIBUTION: Under trees or among the branches of bushes or shrubs in the desert, mostly in flats and washes at 1,000 to 3,500 feet elevation. Arizona Desert. Arizona below the Mogollon Rim from southern Mohave County southward and eastward through all but the lowlands near the Colorado River to Yuma, Gila, and Cochise counties. Mexico in adjacent Sonora.

SYNONYMY: Cereus Greggii Engelm. var. *transmontanus* Engelm. Type locality: "Table Lands of the Gila and San Bernardino," Arizona. *Cereus Greggii* Engelm. var. *roseiflorus* Kuntze. Type locality: Arizona.

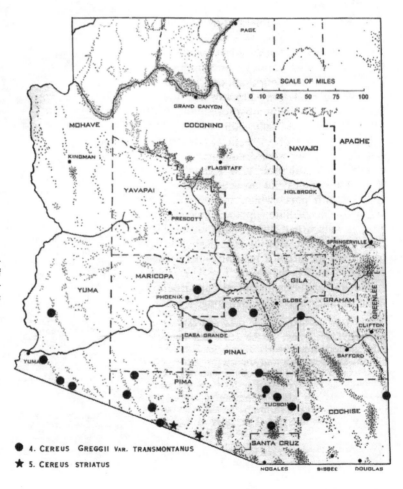

Fig. 2.13. The documented distribution of *Cereus Greggii* var. *transmontanus* and of *Cereus striatus*.

● 4. CEREUS GREGGII Var. TRANSMONTANUS
★ 5. CEREUS STRIATUS

Fig. 2.14. *Cereus striatus.* *A,* stems, and the several tubers, resembling sweet potatoes. *B,* the inconspicuous slender stems above ground. (Photos by A. A. Nichol.)

5. *Cereus striatus*
Brandegee

Suberect plant with clusters of roots each terminating in a tuber 12 to 16 inches long; stems inconspicuous, slender, erect or suberect, 1 to 2½ feet high and about ¼ inch in diameter, gradually enlarged upward from slender bases, branching above; joints elongate and slender; ribs 6 to 9, relatively broad, the intervening grooves narrow; spines appressed, numerous but small, 5 to 10 white ones 1/16 inch long, a few white ones with black tips 1/16 to 1/12 inch long and spreading radially, 2 white ones 1/8 inch long and pointing downward, straight, basally perhaps 1/240 inch in diameter, needlelike, elliptic in cross section; flower nocturnal or diurnal, 2¼ to 3 inches in diameter; superior floral tube more than 1 inch long; petaloid perianth parts white to pink or purple, the largest ones oblanceolate to lanceolate, up to 1¾ inches long, about ⅓ inch broad, mucronate; fruit scarlet at maturity, fleshy, the pulp red, with the surface somewhat spiny, with the spines deciduous, pyriform, about 1½ to 2 inches long, 1 inch in diameter, with the apical cup not prominent, maturing in August, the floral tube persistent; seeds 1/24 inch long.

DISTRIBUTION: Sandy soils of plains and washes; growing in thickets in the desert at 1,500 feet or less elevation. Sonoran Desert. Arizona in Yuma County and on the Papago Indian Reservation near the Mexican Boundary in Pima County. Mexico in Baja California and Sonora.

The species occurs under trees and shrubs in alluvial bottoms and on volcanic hills along the Mexican boundary in Yuma and Pima counties from the Tule Desert to Quijotoa Valley. It is so inconspicuous and so difficult to find when it is not in flower that the range may be considerably more extensive than the areas of its known occurrence. Arizona to Baja California and Sonora.

The seeds are practically all eaten by birds, and the distribution of the species is determined as in *Cereus Greggii.*

The roots are fleshy, light-tan tubers, resembling sweet potatoes. Large plants may produce 2 or 3 dozen tubers.

Commonly the seeds are described as "shining." A recent collection indicates that this may depend upon how much juice of the fruit has dried upon the seed.

SYNONYMY: Cereus striatus Brandegee. *Wilcoxia striata* Britton & Rose. Type locality: San Jose del Cabo, Baja California. *Cereus Diguettii* Weber. *Wilcoxia Diguetii* Diguet & Guillaumin. *Neoevansia Diguetii* W. T. Marshall. *Wilcoxia Diguetii* Peebles. *Peniocereus Diguetti* Backeberg. Type locality: "Basse-Californie . . . aux environs de 27° latitude Nord."

3. ECHINOCEREUS
Hedgehog Cactus

Stems solitary to profusely branching, sometimes 500, the larger ones usually cylindroid, 2 to 24 inches long, 1 to 3 or 4 inches in diameter; ribs 5 to 12, the tubercles coalescent through half to usually nearly all their height. Leaves not discernible in the mature plant. Spines smooth, white, gray, tan, brown, yellow, pink, red, or black; central spines straight or curving, 1/24 to 4 inches long, basally 1/96 to 1/16 inch in diameter, needlelike or flattened; radial spines 3 to 32, straight or curving, ⅛ to 1⅛ inches long, basally 1/96 to 1/24 inch in diameter, needlelike. Flowers and fruits on the old growth of preceding years and therefore located below, usually far below, the apex of the stem or branch, bursting through the epidermis of the stem just above the spine-bearing areole, not quite connected with it, the area of emergence persisting for many years and forming an irregular scar. Flower ¾ to 5 inches in diameter; floral tube above its junction with the ovary funnelform to obconic, green or tinged with the color of the perianth. Fruit fleshy at maturity, with ultimately deciduous areoles bearing spines, globular to ellipsoid, ¼ to 2 inches long, ¼ to 1½ inches in diameter, the floral tube persistent, not regularly dehiscent. Seeds black, obovoid or domelike, reticulate or papillate, longer than broad (length being hilum to opposite side), mostly 1/24 to 1/16 inch long.

DISTRIBUTION: Twenty to thirty species from California to South Dakota and Oklahoma and southward to the vicinity of Mexico City. Eleven species in the United States. Six in Arizona.

LIST OF SPECIES AND VARIETIES

KEY TO THE SPECIES

1. Petaloid perianth parts red or red-and-yellow (with no trace of included blue; probably with plastid pigments); areoles of the mature parts of stems with white felt or cobwebby hairs; flowers not closing at night, open for two or three days until the end of flowering.
    1. *Echinocereus triglochidiatus,* page 121

1. Petaloid perianth parts *either* (1) lavender to purple (probably with betacyanin pigments in solution in the cell sap) *or* (2) yellow (but at least the sepaloid parts yellow with a lavender tinge), green, brownish, or reddish (but then with an admixture of blue); areoles of the *mature vegetative parts of stems not* bearing white felt or cobwebby hairs, the felt of the young areoles persistent one (rarely two) years; flowers closing at night, reopening the next morning (or in hot weather sometimes withering at the end of a single day).

    2. Areoles practically circular.

        3. Principal central spine *not flattened, i.e.,* basally nearly circular to broadly elliptic in cross section, solitary or accompanied by 1 or sometimes 2 or 3 short upper accessory ones, rarely (eastcentral Arizona) only 1/4 to 5/16 inch long and about equal to the accessory centrals and to the longer radials.

            4. Spines *not* yellow or straw-color; principal (lower) central spine straight or curving gently *upward,* tapering gradually from base to apex, the base 1/96 to 1/48 inch in diameter.

                5. Central spine 1, at first at least tipped with dark brown or nearly black, becoming gray in age, curving slightly upward through its entire length or straight; radial spines 7 to 11 per areole, relatively short, ⅜ to ½ inch long, the lowest one longest; stem solitary or with 1 to 5 branches, flabby or firm, usually ovoid to cylindroid-ovoid, 3 to 6 or 10 inches long.
                    2. *Echinocereus Fendleri,* page 129

5. Central spines 2 to 4, the principal one and 1 or sometimes 2 or rarely 3 short accessory ones above it, straight; radial spines 12 to 13 per areole, proportionately elongate, ½ to ¾ inch long; stem at maturity with 3 or usually 5 to 20 branches, firm, cylindroid, elongate, 7 to 12 or 18 inches long.

3. *Echinocereus fasciculatus,* page 132

4. Spines yellow or straw-color; principal central spine basally curved strongly downward, tapering rapidly toward the apex from a broad base about 1/20 inch in diameter; stems elongate-cylindroidal, 10 to 20 inches long.

4. *Echinocereus Ledingii,* page 136

3. Principal spine basally flattened, *i.e.,* narrowly elliptic in cross section, deflexed, the four central spines all well developed, though not necessarily of exactly the same size.

5. *Echinocereus Engelmannii,* page 136

2. Areoles vertically elongate, elliptic to linear, close-set, ¼ to ⅓ inch apart, the spines often obscuring the ribs of the stem; stems usually unbranched or with few branches.

6. *Echinocereus pectinatus,* page 143

1. *Echinocereus triglochidiatus* Engelm.

Plants of most varieties branching several to many times from the bases, each finally forming a dense mound up to 1 foot high and 1 to 4 feet in diameter; joints green or bluish-green, cylindroid to ovoid-cylindroid, 2 to 6 or 12 inches long, 1 to 3 or 4 inches in diameter; ribs 5 to 10 or 12, slightly tuberculate; areoles nearly circular; spines from sparse to dense on the joint, usually gray but sometimes pinkish, straw-color, pale gray, or black, sometimes only 2 or 3 but usually 8 to 12 or 16 per areole, exceedingly variable in the varieties (described further in Table 10); flower 1¼ to 2 inches in diameter, 1½ to 2¼ inches long; petaloid perianth parts red, with no included blue, the largest ones broadly cuneate or cuneate-obovate, ¾ to 1 inch long, ¼ to ½ inch broad, apically rounded and the outer ones slightly mucronulate, entire; fruit red, fleshy at maturity, with a deciduous mass of spines, obovoid to cylindroid, ½ to 1 inch long, ⅜ to ⅝ inch in diameter; seed strongly papillate, 1/16 to 1/2 inch long.

Classification

The complex of local forms composing *Echinocereus triglochidiatus* is deceptive because the extreme types are of radically different appearance. Variation within the species is largely in such striking characters as the size of the stems and the number, size, and angularity of the spines. These variations produce such a diversity of appearance among the extreme types that inevitably "species" have been segregated. These have not been based upon study of natural populations in the field but either (1) upon the few individuals coming into cultivation, usually with no data connecting them with their places in nature, or (2) upon the basis of a single plant or a fragment brought in by an explorer or a traveler or (3) upon knowledge of the flora of a single state or region wherein only one or two or a few forms of the species occur.

Even the relatively small number of herbarium specimens indicates the instability of the populations of the proposed species and their extensive and bewildering intergradation with each other. However, an evaluation can be made only by both extensive and intensive and long-continued study in the field. Twenty-nine years of this has not produced the ultimate answers, but lines of combination and segregation are beginning to emerge.

Although within each of the several varieties some of the natural populations occurring in the field are composed largely of individuals whose character combinations fall within some definable range of variability, even these include a broad spectrum of genetic combinations. In every case investigated in the field, some individuals cannot be included clearly within the variety the general population represents because they have too many characters more abundant in other varieties. Following out the population systems in any direction reveals a gradual loss of some characters and a reduction in the frequency of occurrence of some combinations, there being an irregular trend toward the combination prevalent in other geographically or otherwise segregated varieties. This parallels the trends in combinations of hereditary characters in human populations ranging across Europe into and across Asia and Africa.

The central type in the complex is var. *melanacanthus,* the character combinations of which can be discerned in at least some plants through the entire central range of the species and at many points on the periphery — from California to Colorado, Texas, and Mexico as far south as Durango. This variety is in contact at one point or another with each of the others, and it intergrades with all of them. This is a common pattern within a species, for evolution may proceed rapidly in local areas where on the margins of distribution usually ecological conditions are critical and where a small breeding population is shielded from the general population. Differences thus become locally partly stabilized, and varieties and ultimately species may result.

Var. *melanacanthus* does not appear in accounts of the flora of California, but one collection is clearly of this variety, and numerous plants not or barely distinguishable from it are to be found in the populations of var. *mojavensis* in the hills above the Mojavean Desert from California to southwestern Utah and northwestern Arizona. There is a clear trend toward var. *melanacanthus* from the extreme and relatively local var. *triglochidiatus* (mostly in New Mexico) through var. *gonacanthus* (Colorado, the eastern edge of northern Arizona, and northwestern and northern New Mexico). Intergradation of var. *gonacanthus* with vars. *melanacanthus* and *neomexicanus (polyacanthus)* is common in New Mexico.

Nomenclature

Echinocereus triglochidiatus has a complex nomenclatural history. It was named as a species first in 1848 — five times within thirty-three days. The nomenclatural combinations, each applied to some element of the species as interpreted here, were *Echinocereus triglochidiatus* Engelm. and *Echinocereus coccineus* Engelm., January 13; *Cereus Reomeri* Mühlenpfordt and *Echinopsis octacantha* Mühlenpfordt, January 15; and *Mammillaria aggregata* Engelm., February 17. Fortunately the dates of publication are clear, though establishing the identity of three of the proposed species is complex. However, these are the three later names, the first two being of clear application.

Echinocereus triglochidiatus and *Echinocereus coccineus,* applied to the extremes of the complex, were described and published simultaneously by Engelmann, and either name combination might have been chosen for the entire complex of intergrading forms and varieties. However, according to the "rule of the first reviser," the first author treating two or more simultaneously published species (or other taxa) as subdivisions of a single species (or other taxon) or as synonyms must be followed in his choice of one of the names for the combined species (taxon) (International Code of Botanical Nomenclature, 1966, Article 57). In 1941, W. T. Marshall wisely treated the two entities mentioned above as varieties of a single species, and he published the combination *Echinocereus triglochidiatus* var. *coccineus* for one of them. Although, especially in the light of new information, the choice for specific rank was unfortunate, *coccineus* being a shorter epithet and being based upon a much more common and widespread variety (and the one intergrading with all the others), the decision of long ago must be followed. Otherwise there could be no stability of nomenclature.

In choosing the epithet *coccineus* to designate the variety, earlier application of several epithets of varietal rank was overlooked. Despite the earlier (1848) publication of *coccineus* in specific rank, all these had priority in varietal rank. *The rule is to adopt the epithet first published in the proper rank.* "In no case does a name or epithet have priority outside its own rank" (Article 60). Var. *melanacanthus* Engelm. was published in 1849, and in varietal rank the epithet has 92 years priority over *coccineus.* Upon the basis of field and herbarium evidence, the two epithets were based upon type specimens collected from plants bearing the same relationships as a blue-eyed man to his brown-eyed mother. Use of *melanacanthus* is mandatory under the Code if this population system is treated as a variety under any species of any genus.

Some have questioned the applicability of the epithet *melanacanthus* (black-spined) because ordinarily the spines are gray to pink, tan, or straw-colored. Nearly everywhere in the field the spine color varies greatly from individual to individual. Occasionally, about the type locality, Santa Fé, there is a plant with black or very dark brown spines. However, this feature is not correlated with any other, and classification of taxa of any rank is not by single characters but by consistently occurring combinations of characters. Collections of plants with black spines and of others with the more common spine colors were made in the same population in a canyon 15 miles southeast of Santa Fé at 7,400 feet elevation *(L. Benson, 14,694, Pom, 14,694a, Pom)*. Under the Code the question of whether an epithet is truly descriptive of a species or a variety or of a form common within it is of no consequence. "The purpose of giving a name to a taxon is not to indicate its characters or history, but to supply a means of referring to it." "A legitimate name or epithet must not be rejected merely because it is inappropriate or disagreeable, or because another is preferable or better known, or because it has lost its original meaning" (Article 62). The only question is whether the epithet was applied to *any element whatever* within the system of populations forming the taxon, and this is shown by the type specimen. Whether or not this element is predominant or even common is beside the point.

Although the epithet *melanacanthus* has long priority over *coccineus* in varietal rank, the reverse is true in specific rank, which was accorded to *coccineus* in 1848 and to *melanacanthus* only in 1963, after a lapse of 115 years. The combination *Echinocereus melanacanthus* was attributed in a popular book to Engelmann by an author who accepted the proposition that *coccineus* and *melanacanthus* are the same, as shown by retention of *Echinocereus coccineus* in the list of synonyms under *melanacanthus*. Perhaps this was because, after Engelmann in 1849 combined *Echinocereus* with the earlier-published genus *Cereus,* in 1856 he abandoned *coccineus* and substituted a *nomen novum, Cereus phoeniceus*. This was because he had found the combination *Cereus coccineus* to be preoccupied by an earlier usage in 1828 by Salm-Dyck for another species. However, the inavailability of a place for the epithet *coccineus* for a species under *Cereus* did not affect its usage under *Echinocereus* or any other genus.

1A. Var. *melanacanthus*
     (Engelm.) L. Benson
     (Table 10)

DISTRIBUTION: Rocky hillsides, ledges, and canyons; mostly on igneous rocks at 3,500 to 8,000 or 9,600 feet elevation. Rocky Mountain Montane Forest, Juniper-Pinyon Woodland, and the edges of the Desert Grassland and the Great Plains Grassland. Arizona, except in the lower desert regions of

Fig. 3.1. A red-flowered hedgehog cactus, *Echinocereus triglochidiatus* var. *melanacanthus*. (Photo by A. A. Nichol.)

TABLE 10. CHARACTERS OF THE VARIETIES OF ECHINOCEREUS TRIGLOCHIDIATUS

	A. Var. melanacanthus	B. Var. mojavensis	C. Var. neomexicanus	D. Var. arizonicus	E. Var. gonacanthus	F. Var. triglochidiatus
Stem number	Ultimately numerous, up to 500.	Ultimately numerous, up to 500.	Mostly 5 to 45.	Few.	Few.	Few.
Stem length	1½ to 3 or 6 inches.	1½ to 3 or 6 inches.	8 to 12 inches.	9 inches.	3 to 5 inches.	6 to 12 inches.
Stem diameter	1 to 2 or 2½ inches.	1 to 2 or 2½ inches.	3 to 4 inches.	6 to 10 inches.	2 to 3 inches.	Mostly about 3 inches.
Stem ribs	Mostly 9 or 10, tuberculate.	Mostly 9 or 10, tuberculate.	8-12, mostly 10, not markedly tuberculate.	About 10, tuberculate.	About 8, tuberculate.	5-8, tuberculate.
Spines	Gray, black, pink, or basally tan, or sometimes straw-color, up to 1 to 2½ inches long, nearly straight, rarely angled.	Gray, pink, or at first straw-color, usually up to 1¾ to 2¾ inches long, striate, smooth or angled.	Tan or pink, becoming light gray, up to 1½ inches long, nearly straight, not angled.	Dark gray, up to 1 to 1½ inches long, nearly straight, not angled.	Gray or tan, 1 to 1¾ inches long, nearly straight, 6 or (3-4), 4-angled.	Gray, ¾ to 1 inch long, nearly straight, 3-angled.
Central spines	1-3, light or dark, spreading or the longest deflexed, up to 1/32 inch in basal diameter.	1-2, light, usually twisting, often striate, about 1/32 inch in basal diameter.	2-4, gray, spreading, 1/48 to 1/24 inch or a little more in basal diameter.	1-3, the largest deflexed, acicular, gradually tapering, with minute striations, up to 1/16 inch in basal diameter.	1 (or 0-2), gray, spreading, up to twice as long as the radial, up to 1/20 inch thick, 6-7-angled.	0 (or rarely 1 and then like the radial).
Radial spines	5-11, half as long to sometimes nearly as long as the central.	5-8, half as long to sometimes nearly as long as the central.	9-12, tannish or light gray, about half as long as the central.	5-11, often slightly curved, pinkish-tan, shorter than the central.	5-8, tan or gray, up to 1/24 inch in diameter	3-6, tan or gray, spreading or recurving, up to 1/16 inch thick.
Flower shape & approximate size	Slender, 1 to 1½ inches in diameter, 1¼ to 2 or 2½ inches long.	Slender, 1½ to 2 inches in diameter, 1½ to 2 inches long.	Slender, 1½ inches in diameter, 2 to 2¾ inches long.	Broad, about 2 inches in diameter, 2½ inches long.	Broad, 2¾ inches in diameter, 2½ inches long.	Broad, 2 inches in diameter, 2 to 2½ inches long.
Style (approximate size)	1/24 inch in diameter, equal to or longer than the perianth.	1/24 inch in diameter, equal to or longer than the perianth.	1/24 inch in diameter, about equal to or longer than the perianth.	1/12 inch in diameter, equal to the perianth.	..	..
Geographical distribution	Upland Arizona. Central Utah to southern Colorado and southwestern Texas. Southward in Mexico to Durango.	Northwestern Arizona in Mohave County. South-eastern California; southern Nevada; south-western corner of Utah.	Southeastern Arizona. Southwestern and south-central New Mexico; Trans-Pecos Texas. Northwestern Mexico.	Arizona between Superior and Globe.	Northern edge of Arizona. Southcentral and south-western Colorado; north-ernmost New Mexico.	Near Ft. Defiance, Arizona. Southernmost Colorado; westcentral and central New Mexico.

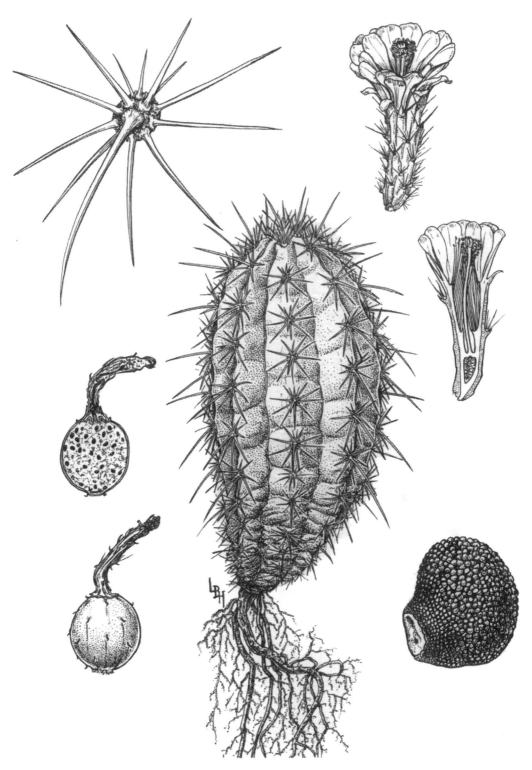

Fig. 3.2. A red-flowered hedgehog cactus, *Echinocereus triglochidiatus* var. *melanacanthus*. *Upper left*, spine cluster and areole. *Upper right*, flower in external view and in longitudinal section. *Lower left*, fruit in side view and longitudinal section. *Below*, young stem and root system. Note the clusters of old stems in Fig. 3.1 and Fig. 3.3, *above*. *Lower right*, seed.

Yuma, Maricopa, western Pinal, and western Pima counties. California on Clark Mountain, Mojave Desert region; Nevada in the Antelope Mountains, Eureka County; Utah from Juab, Carbon, and Uintah counties southward; Colorado from the Colorado River Drainage to Pueblo County and southward; New Mexico in all the higher areas, but not on the Great Plains or in the deserts; Texas in the Trans-Pecos region and in granitic areas of the Edwards Plateau and its escarpment as far eastward as Granite Mountain, Burnet County. Mexico in the region of the Sierra Madre Occidental as far south as Durango.

The only collection from Yuma and Maricopa counties, Arizona, is from Ft. Harquehala.

Variations in characters of the spines are common. Long- and short-spined individuals grow in the same natural populations, and various intergrading forms occur with them. Some individuals may have striations or angles on basal portions of the longest spines. Black spines (sources of the epithet of the variety) are rare, but plants having them do occur, for example, along with others having the usual spine colors, in New Mexico near the type locality of var. *melanacanthus*. Plants with only yellow, long spines occur in west-central Utah.

This variety intergrades with all the others, and intermediate plants are common along the zones of geographical contact.

SYNONYMY: Echinocereus coccineus Engelm. *Cereus coccineus* Engelm. *Cereus phoeniceus* Engelm., *nom. nov.* for *Echinocereus coccineus* Engelm., this being necessary under *Cereus. Echinocereus phoen[i]ceus* Engelm. ex Rümpler. *Echinocereus triglochidiatus* Engelm. var. *coccineus* Engelm. ex. W. T. Marshall. Type locality: Santa Fé, New Mexico. *Cereus Roemeri* Mühlenpfordt, Jan. 15, 1848, not Engelm. in 1849. *Echinocereus Roemeri* Rydb. Neotype locality: Fredericksburg, Gillespie County, Texas. *Echinopsis octacantha* Mühlenpfordt. *Cereus octacanthus* Coulter. *Echinocereus octacanthus* Britton & Rose. *Echinocereus coccineus* Engelm. var. *octacanthus* Boissevain. *Echinocereus triglochidiatus* Engelm. var. *octacanthus* Mühlenpfordt ex W. T. Marshall. No locality or specimen mentioned. *Mammillaria aggregata* Engelm. *Cereus aggregatus* Coulter. *Echinocereus aggregatus* Rydb. *Coryphantha aggregata* Britton & Rose. *Mammillaria vivipara* (Nutt.) Haw. var. *aggregata* L. Benson. *Escobaria aggregata* F. Buxbaum. *Coryphantha vivipara* (Nutt.) Britton & Rose var. *aggregata* W. T. Marshall. Type locality: above San Lorenzo, New Mexico. *Cereus Roemeri* Engelm. in 1849, not Mühlenpfordt in 1848. *Echinocereus Roemeri* Engelm. ex Rümpler. Type locality: Llano River, Texas. *Cereus coccineus* (Engelm.) Engelm. var. *melanacanthus* Engelm. *Echinocereus triglochidia-tus* Engelm. var. *melanacanthus* L. Benson. *Echinocereus melanacanthus* Engelm ex. W. H. Earle. Type locality: Santa Fé, New Mexico. *Cereus coccineus* (Engelm). Engelm. var. *cylindricus* Engelm. Type locality: Santa Fé, New Mexico. *Cereus hexaedrus* Engelm. *Echinocereus hexaedrus* Engelm. ex Rümpler. *Echinocereus paucispinus* (Engelm.) Engelm. ex Rümpler var. *hexaedra* [*hexaedrus*] K. Schum. *Echinocereus triglochidiatus* Engelm. var. *hexaedrus* Boissevain. Type locality: Near Zuñi, New Mexico. *Cereus conoideus* Engelm. & Bigelow. *Cereus phoeniceus* Engelm. ssp. *conoideus* Engelm. *Echinocereus phoeniceus* Rümpler var. *conoideus* K. Schum. Type locality: Anton Chico, upper Pecos River, New Mexico. *Cereus mojavensis* Engelm. & Bigelow var. *zuniensis* Engelm. & Bigelow. *Cereus Bigelovii* Engelm. var. *zuniensis* Engelm. *Echinocereus mojavensis* (Engelm.) Engelm. var. *zuniensis* Engelm. ex Rümpler. Type locality: Canyon Diablo east of Flagstaff, Arizona. *Echinocereus Krausei* De Smet. "Vaterland unbekannt." *Echinocereus phoeniceus* (Engelm.) Engelm. ex Rümpler var. *inermis* K. Schum. Type locality: "Herr C. A. Purpus aus Colorado. . . ." *Echinocereus canyonensis* Clover & Jotter. Type locality: Bass Cable below Hermit Creek Rapids, Grand Canyon, Coconino County, Arizona.

Plant of Uncertain Identity and Relationship

Echinocereus Kunzei Gürke. *Echinocereus coccineus* Engelm. var. *Kunzei* Backeberg. Type locality: "Phoenix in Arizona." No specimen was mentioned.

Possible Variety

The specimens discussed here cannot be categorized satisfactorily at present. They may represent a variety of *Echinocereus triglochidiatus* near var. *melanacanthus*, but this is not certain. Field work is needed to determine the nature of local populations.

Stems up to 16 or 17 inches long and only about 1 inch in diameter; spines brown, 9 or 10 per areole, ½ to 1 inch long, very slender, 1/120 to 1/72 inch in diameter.

SYNONYMY: Echinocereus decumbens Clover & Jotter. *Echinocereus Engelmannii* (Parry) Lemaire var. *decumbens* L. Benson. Type locality: limestone ledge 30 feet from the Colorado River's edge at the base of rocky talus at Mile 16½ Marble Canyon, Coconino County, Arizona. A similar collection from Coconino County, Arizona, is "Rock ledge below Mooney Falls."

1B. Var. *mojavensis*
    (Engelm.) L. Benson
    (Table 10)

DISTRIBUTION: Rocky hillsides and canyons in woodland above the deserts at

Fig. 3.3. Red-flowered hedgehog cacti, *Echinocereus triglochidiatus. Above,* var. *melana-canthus* in flower. *Below,* var. *arizonicus; left,* flower, showing the rupture of the epidermis of the stem just above the spine-bearing areole, where the developing flower bud forced its way outward, this being a unique character of *Echinocereus; right,* stem and flower. The red flower color includes no admixture of blue. It is due to a plastid pigment not soluble in water. See Fig. 3.11 for the lavender to magenta flower color of the other species, including *Echinocereus fasiculatus.*

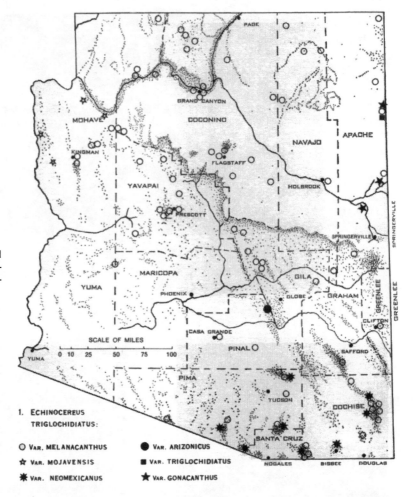

Fig. 3.4. The documented distribution of *Echinocereus triglochidiatus,* according to its varieties.

1. ECHINOCEREUS
   TRIGLOCHIDIATUS:

○ Var. MELANACANTHUS   ● Var. ARIZONICUS

☆ Var. MOJAVENSIS   ■ Var. TRIGLOCHIDIATUS

✸ Var. NEOMEXICANUS   ★ Var. GONACANTHUS

3,500 to 8,000 or 10,000 feet elevation. Juniper-Pinyon Woodland and the California Chaparral (desert-edge phase) or Rocky Mountain Montane Forest. Arizona in western Mohave County (according to the map prepared by A. A. Nichol, in Benson *et al., The Cacti of Arizona,* ed. 1, 1940; ed. 2, 1950, but this not so far confirmed by specimens). California from Inyo County southward to the desert side of the San Jacinto Mountains and eastward to the boundary of Nevada; southcentral and southern Nevada; Utah in Washington County.

Even in its extreme populations this variety includes some plants which have in greater or lesser degree, the characters of var. *melanacanthus,* and the reverse is true in the populations of var. *melanacanthus* along the zone of contact with that variety in Washington County, Utah. In the Charles-

ton Mountains, Nevada, the characters of var. *mojavensis* are more abundant, but those of var. *melanacanthus* are proportionately more common in the populations than in California.

SYNONYMY: Cereus mojavensis Engelm. & Bigelow. *Cereus Bigelovii* Engelm., probably a discarded name intended to have been replaced by *Cereus mojavensis. Echinocereus mojavensis* Engelm. ex Rümpler. *Echinocereus triglochidiatus* Engelm. var. *mojavensis* L. Benson. Type locality: Mojave River, California. *Echinocereus Sandersii* Orcutt. Type locality: Fenner, California.

1C. Var. *neomexicanus*
      (Standley) Standley ex W. T. Marshall
      (Table 10)

DISTRIBUTION: Soils of igneous origin in woodlands and grasslands at 4,500 to 7,000 feet elevation. Southwestern Oak Woodland and Chaparral and the oak woodlands of western Texas and northwestern

Mexico; Juniper-Pinyon Woodland. Arizona from Pima and Santa Cruz counties to Cochise County. New Mexico from McKinley County to Bernalillo, Hidalgo, Chaves, and Eddy counties; Texas west of the Pecos River. Mexico from Chihuahua to Durango.

SYNONYMY: Echinocereus polyacanthus Engelm. *Cereus polyacanthus* Engelm. *Echinocereus triglochidiatus* Engelm. var. *polyacanthus* L. Benson. Type locality: Cosihuiriachi, Chihuahua, Mexico. *Echinocereus neo-mexicanus* Standley. *Echinocereus triglochidiatus* Engelm. var. *neomexicanus* Standley ex W. T. Marshall. Type locality: Organ Mountains, Doña Ana County, New Mexico. *Echinocereus Rosei* Wooton & Standley. *Echinocereus triglochidiatus* Engelm. var. *Rosei* W. T. Marshall. Type locality: Agricultural College, New Mexico.

1D. Var. **arizonicus**
(Rose ex Orcutt) L. Benson
comb. nov. (p. 21)
(Table 10)

DISTRIBUTION: Chaparral areas in the mountains at about 3,500 to 4,700 feet elevation. Southwestern Oak Woodland and Chaparral. Arizona in the mountainous area near the line between Gila and Pinal counties.

SYNONYMY: Echinocereus arizonicus Rose ex Orcutt. Type locality: near boundary monument between Pinal and Gila counties, Arizona.

1E. Var. *gonacanthus*
(Engelm. & Bigelow) Boissevain
(Table 10)

DISTRIBUTION: Rocky hillsides in or near woodlands at 5,600 to 7,900 feet elevation. Juniper-Pinyon Woodland. Arizona in easternmost Apache County. Colorado from Chaffee and Fremont counties to Archuleta and Rio Grande counties; western New Mexico sparingly from San Juan and Rio Arriba counties to Grant and Socorro counties and in forms transitional to vars. *melanacanthus* and *neomexicanus* eastward as far as Colfax and Eddy counties. This variety occupies a range northwest and north of that of var. *triglochidiatus,* with which it also intergrades.

SYNONYMY: Cereus gonacanthus Engelm. & Bigelow. *Echinocereus gonacanthus* Engelm. ex Rümpler. *Echinocereus paucispinus* (Engelm.) Engelm. ex Rümpler. var. *gonacanthus* K. Schum. *Echinocereus triglochidiatus* Engelm. var. *gonacanthus* Boissevain. *Echinocereus triglochidiatus* Engelm. var. *gonacanthus* Engelm. & Bigelow ex W. T. Marshall. Type locality: Apache County, Arizona, near Zuñi, New Mexico.

1F. Var. *triglochidiatus*
(Table 10)

DISTRIBUTION: Rocky or gravelly ridges, hills, and canyons in woodlands at 4,350 to 6,900 feet elevation. Juniper-Pinyon Woodland. Arizona near Ft. Defiance, Apache County (probably the population transitional to var. *gonacanthus*). Southernmost Colorado near Arboles, Archuleta County (there in a population transitional to var. *gonacanthus*); New Mexico.

SYNONYMY: Echinocereus triglochidiatus Engelm. *Cereus triglochidiatus* Engelm. *Echinocereus paucispinus* Engelm. var. *triglochidiatus* K. Schum. Type locality: Wolf Creek, New Mexico.

2. *Echinocereus Fendleri*
Engelm.

Mature plants with usually 1 but up to 5 stems, these flabby or firm, the clumps not dense; larger stems green, ovoid to ovoid-cylindroid or cylindroid, 3 to 6 or 10 inches long, 1½ to 2½ inches in diameter; ribs 8 or usually 9 or 10, not markedly tuberculate; areoles circular; spines not dense, not obliterating the stem surface; central spine very dark or pale gray and tipped with brown or black at first but changing to gray in age, solitary in the areole, curving slightly upward through its entire length or straight and at right angles to the stem, rigid, ½ to 1½ inches long, basally 1/48 to 1/36 inch in diameter, needlelike, circular in cross section, tapering gradually; radial spines usually white or pale gray, sometimes yellow or like the central spine, 7 to 11 per areole, spreading at a low angle, straight, short, the lowest one longer than the others, the longer ones ⅜ to ½ inch long, basally up to 1/96 to 1/48 inch in diameter; flower 2 to 2½ inches in diameter; petaloid perianth parts magenta, the largest ones elliptic-cuneate, 1¼ to 1½ inches long, about ½ inch broad, acute, mucronulate entire; fruit green, turning reddish, ¾ to 1¼ inches long, ½ to 1 inch in diameter; seed 1/16 inch long.

2A. Var. *Fendleri*
Stems usually ovoid to ovoid-cylindroid but in old plants sometimes cylindroid, 3 to 6 or 10 inches long; central spine very dark at first but changing to gray in age, curving upward through its entire length, 1 to 1½

Fig. 3.5. A hedgehog cactus, *Echinocereus Fendleri*, with a withered flower or young fruit. (Photograph by Robert H. Peebles.)

inches long; radial spines ⅜ to ½ inch long, 1/96 to 1/48 inch in diameter, all straight.

DISTRIBUTION: Sandy or gravelly areas in grasslands and woodlands at mostly 6,000 to 8,000 feet elevation. Great Plains Grassland, Juniper-Pinyon Woodland, and the lower edge of the Rocky Mountain Montane Forest. Arizona from Coconino County eastward and southward above the Mogollon Rim and sparingly southward at higher levels to Santa Cruz and Cochise counties. Utah in the Henry Mountains, Garfield County; southern Colorado (occasional); New Mexico; Texas in El Paso County. Mexico in northern Chihuahua.

SYNONYMY: Cereus Fendleri Engelm. *Echinocereus Fendleri* Engelm. ex. Rümpler. Type locality: Santa Fé, New Mexico. *Cereus Fendleri* Engelm. var. *pauperculus* Engelm. Type locality: Santa Fé, New Mexico.

2B. Var. *rectispinus*
(Peebles) L. Benson

Stems cylindroid, 4 to 7 or 10 inches long; central spine pale gray but at least tipped with brown or black, straight and projecting at right angles to the stem, ½ to 1½ inches

Fig. 3.6. Hedgehog cacti, *Echinocereus Fendleri*. A, var. *rectispinus*, in flower and B, var. *Fendleri*, in flower. (Photographs by Robert H. Peebles.)

long; radial spines 3/8 to 1/2 inch long, 1/96 to 1/48 inch in diameter, all straight.

DISTRIBUTION: Sandy or gravelly soils in grassland at mostly 3,900 to 5,500 or 6,800 feet elevation. Desert Grassland. Southeastern Arizona from the Santa Catalina Mountains in Pinal and Pima counties to near Clifton in Greenlee County and eastward and southward to Graham, Santa Cruz, and Cochise counties. Southwestern New Mexico from Grant County to Hidalgo and Sierra counties; Texas from El Paso County to Culberson County.

SYNONYMY: Echinocereus rectispinus Peebles. *Echinocereus Fendleri* Engelm. var. *rectispinus* L. Benson. *Echinocereus fasciculatus* (Engelm.) L. Benson var. *rectispinus* (Peebles) L. Benson. Type locality: near Nogales, Arizona.

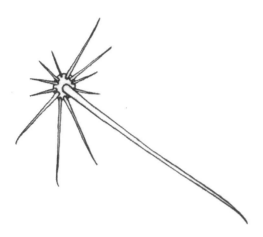

Fig. 3.7 *Echinocereus fasciculatus* var. *rectispinus.* Spine cluster with the single central spine at right angles to the stem.

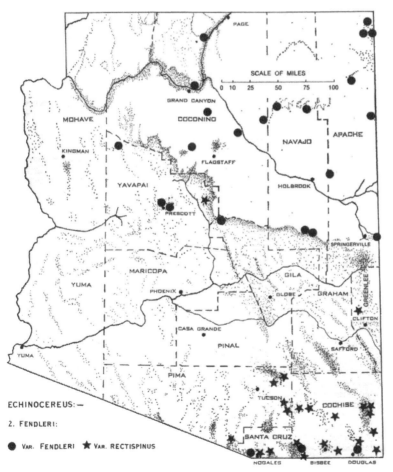

ECHINOCEREUS: —

2. FENDLERI:

● VAR. FENDLERI ★ VAR. RECTISPINUS

Fig. 3.8. The documented distribution of *Echinocereus Fendleri,* according to its varieties.

3. Echinocereus fasciculatus
(Engelm.) L. Benson, **comb. nov.** (p. 21)

Stems 3 or 5 to 20, the tissues firm, not flabby, the clumps not dense; stems green, elongate-cylindroid, 7 to 12 or 18 inches long, 1½ to 2½ or 3 inches in diameter; ribs 8 to 18, not markedly tuberculate; areoles circular; spines not dense, not obliterating the stem; central spines either pale gray, straw-color, white, light brown, tan, or reddish-brown, the principal one conspicuous, spreading or turned downward, straight, accompanied by 1 or 2 or infrequently 3 short accessory centrals, 1 to 3 inches (except in var. *Bonkerae* only 1/4 to 5/16 inch) long, basally 1/48 to 1/36 or 1/24 inch in diameter, needlelike, circular in cross section, tapering gradually; radial spines white or pale gray, sometimes yellow or like the central spine, 12 to 13 per areole, spreading at a low angle, straight, mostly 1/2 to 3/4 inches long, basally up to 1/48 inch in diameter, needle-like; flower 2 to 2½ inches in diameter; petaloid perianth parts magenta to reddish-purple, the largest ones elliptic-cuneate, 1¼ to 1½ inches long, about ½ inch broad, acute, mucronulate, entire; fruit green, turning reddish, ¾ to 1¼ inches long, ½ to 1 inches in diameter; seed 1/16 inch long.

3A. Var. *fasciculatus*
(Table 11)

DISTRIBUTION: Sand, gravel, or rocks of hills and washes in the desert at 2,500 to 3,500 or 5,000 feet elevation. Arizona Desert and barely in the Desert Grassland. Arizona from Yavapai County and Oak Creek Canyon in Coconino County southward through the higher desert near the Mogollon Rim to Gila County and southwestward to Pima and Cochise counties. New Mexico in Grant County (rare) and Hidalgo County.

SYNONYMY: Mammillaria fasciculata Engelm. *Cactus fasciculatus* Kunze. *Neomammillaria fasciculata* Britton & Rose. *Ebnerella fasciculata* F. Buxbaum. Gila River below the mouth of Bonanza Creek, Graham County, Arizona. *Echinocereus rectispinus* Peebles var. *robustus* Peebles. *Echinocereus robustus* Peebles. *Echinocereus Fendleri* Engelm. var. *robustus* L. Benson. Type locality: Tucson to Sabino Canyon, Pima County, Arizona.

3B. Var. **Boyce-Thompsonii**
(Orcutt) L. Benson, **comb. nov.** (p. 21)
(Table 11)

DISTRIBUTION: Gravelly or rocky slopes in the desert at about 1,000 or 2,000 to 3,000 feet elevation. Arizona Desert. Arizona below the Mogollon Rim from Yavapai County to eastern Maricopa, Gila, and northeastern Pinal counties.

A specimen from the Apache Trail near Fish Creek Canyon has principal central spines up to 4 inches long. The two accessory central spines are up to 1⅛ inches long.

SYNONYMY: Echinocereus Boyce-Thompsonii Orcutt. *Echinocereus Fendleri* (Engelm.) Lemaire var. *Boyce-Thompsonii* L. Benson. Type locality: Boyce-Thompson Southwestern Arboretum, near Superior, Arizona.

TABLE 11. CHARACTERS OF THE VARIETIES OF ECHINOCEREUS FASCICULATUS

	A. Var. fasciculatus	B. Var. **Boyce-Thompsonii**	C. Var. **Bonkerae**
Stems	5-20, 7 to 18 inches long, cylindroid; ribs 8-10.	4-12, 4 to 10 inches long, cylindroid; ribs 12 or usually 14-18 or more.	Usually 5-15, 5 to 8 inches long, cylindroid, ribs 11 or 13-16.
Central spines	Pale gray but at least tipped with brown or black, standing at right angles to the stem or deflexed, straight, somewhat flexible, 1 to 3 inches long.	Principal one light-colored, either straw, reddish-tan, or white and light brown, straight, deflexed, flexible, very slender, 1½ to 2 or 4 inches long.	White or pale gray tipped with brown, standing at right angles to the stem, stout, straight, rigid, **only ¼ to 5/16 inch long.**
Geographical distribution	Arizona Desert. Arizona from Yavapai County to Pima and Graham counties and rare in Cochise County. New Mexico in Hidalgo County.	Arizona Desert. Arizona from Yavapai County to Gila County.	Desert Grassland. Arizona in Santa Cruz, Pinal, Gila, Graham, and northwestern Cochise counties.

Fig. 3.9. A hedgehog cactus, *Echinocereus fasciculatus*. (Photographs: *above,* by J. J. Thornber; *below,* by Robert H. Peebles.)

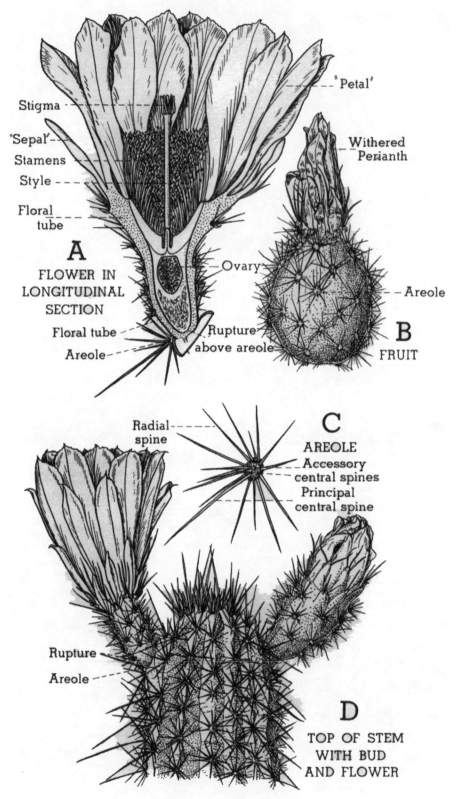

Fig. 3.10. A hedgehog cactus, *Echinocereus fasciculatus*.

Fig. 3.11. Hedgehog cacti, *Echinocereus fasciculatus*. *Above*, var. *fasciculatus* in flower. Note the gray spines. The flowers are in the color series from lavender to purple, a mixture of red and blue. The betacyanin pigments are water soluble. Compare the flower color in Fig. 3.3. *Below*, var. *Boyce-Thompsonii*, showing the elongate and numerous yellow spines.

Fig. 3.12. A hedgehog cactus, *Echinocereus fasciculatus* var. *Boyce-Thompsonii*. Stem with a red fruit from which the spine clusters have fallen. (Photograph by Robert H. Peebles.)

3C. Var. **Bonkerae**
(Thornber & Bonker) L. Benson
comb. nov. (p. 22)
(Table 11)

DISTRIBUTION: Sandy or gravelly soils of plains, hills, and canyons in grassland at 3,000 to 5,000 or 6,000 feet elevation. Desert Grassland. Arizona from Gila County to Pinal, Graham, and northwestern Cochise counties and near Nogales, Santa Cruz County.

Britton & Rose (Cactaceae 2:37. *f. 45.* 1922) published a photograph of this plant as *Echinocereus Fendleri.*

SYNONYMY: Echinocereus Bonkerae Thornber & Bonker. *Echinocereus Fendleri* Engelm. var. *Bonkerae* L. Benson. *Echinocereus Boyce-Thompsonii* Orcutt var. *Bonkerae* Peebles. Type locality: near Oracle at the north base of the Santa Catalina Mountains, Pinal County, Arizona.

4. *Echinocereus Ledingii*
Peebles

Stems usually 4 to 10, the clumps not dense, the stems green, the tissues firm, ovoid to cylindroid, often elongate, 10 to 20 inches long, commonly 1½ to 3 inches in diameter; ribs usually 12 to 14 or 16, not markedly tuberculate; areoles circular; spines yellow or straw-color (as with other yellow spines, turning black in age or in pressed specimens); central spine solitary in the areole or accompanied by 1 to 3 short upper centrals, the principal one strongly curved near the base and turned downward, stout and rigid, 3/4 to 1 inch long, basally very thick and 1/20 inch in diameter, needlelike, circular in cross section, tapering rapidly from the broad base; radial spines about 9 to 11 per areole, spreading at a low angle, the longer ones 1/2 to 5/8 inches long, basally about 1/48 inch in diameter, needlelike; flower 2 to 2½ inches in diameter; petaloid perianth parts magenta to rose-purple, the largest elliptic-cuneate, 1¼ to 1½ inches long, about ½ inch broad, acute, mucronulate, entire; fruit green, turning reddish, 3/4 to 1 1/4 inches long, 1/2 to 1 inch in diameter, seed 1/16 inch long.

DISTRIBUTION: Sandy or gravelly mountain slopes in grassland, woodland, or chaparral at 4,000 to 6,000 feet elevation. Southwestern Oak Woodland and Chaparral and the edge of the Desert Grassland. Arizona: known from specimens from the Santa Catalina Mountains in Pima County, the Graham or Pinaleno Mountains in Graham County, and the Mule Mountains in Cochise County; reported by A. A. Nichol to occur also in the Quinlan, Santa Rita, Huachuca, and Chiricahua mountains (*cf.* in L. Benson, *Cacti of Arizona.* ed. 2. 89. *f. 23. upper left,* 1950).

SYNONYMY: Echinocereus Ledingii Peebles. Type locality: Mt. Graham, Graham County, Arizona.

5. *Echinocereus Engelmannii*
(Parry) Lemaire

Stems 5 to 60, open or in dense masses, or mounds, clumps up to 1 or 2 feet high and 2

Fig. 3.13. A hedgehog cactus, *Echinocereus fasciculatus* var. *Bonkerae.* (Photographs: *above,* by A. A. Nichol; *below* by J. J. Thornber.) For var. *Bonkerae,* see also the dust jacket.

or 3 feet in diameter; stems green, cylindroid, usually elongate, 6 to 24 inches long, 1¼ or usually 2 to 3 inches in diameter; ribs commonly 10 to 13, the tubercles not prominent; areoles circular; spines numerous but not obliterating the surface of the stem; central spines variable (*cf.* Table 12), 4 to 6 per areole, the lower (principal) one declined and flattened, the others variable (Table 12), straight or curved or twisted, the longer ones 1 to 2½ inches long, basally 1/25 to 1/16 inch broad, flattened, narrowly elliptic in cross section; radial spines variable (Table 12) 6 to 12 per areole, spreading close to the stem, straight, the longer ones mostly ½ to 1 inch long, basally averaging about 1/48 inch broad, acicular, elliptic in cross section; flower 2 to 3 inches in diameter and of the same length; petaloid perianth parts purple to magenta or lavender, the larger ones cuneate-oblanceolate, 1½ to 2 inches long, up to nearly 1 inch broad, rounded, mucronulate; fruit green, turning to red at maturity, with the spine clusters deciduous, ¾ to 1¼ inches long, about ½ to 1 inch in diameter; seed 1/16 inch long.

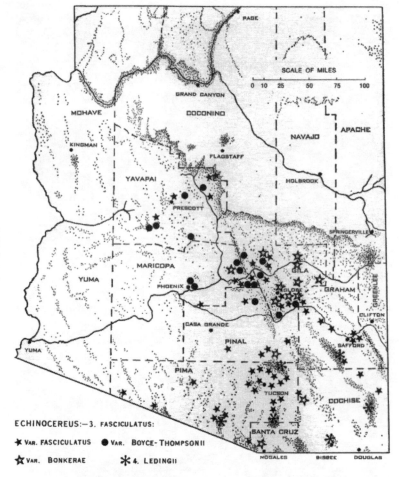

Fig. 3.14. The documented distribution of *Echinocereus fasciculatus,* according to its varieties, and of *Echinocereus Ledingii.*

ECHINOCEREUS:—3. FASCICULATUS:

★ VAR. FASCICULATUS   ● VAR. BOYCE-THOMPSONII

☆ VAR. BONKERAE   ✳ 4. LEDINGII

The red fruits are edible, and being plentiful, they are important as food for birds and rodents. The ovary is rich in sugar, and the seeds contain an abundance of fat. The Pima Indians at Sacaton consider the fruits to be a delicacy.

5A. Var. *Engelmannii*
   (Table 12)

DISTRIBUTION: Gravel, sand, or rocks of hills, washes, and canyons in the desert at 2,000 to 5,000 feet elevation. Mojavean and Arizona but principally the Colorado Desert. Arizona (rare) in Yuma, Maricopa, Pima, and Cochise counties. Califorina in the Mojave Desert and in the mountains of the northern and western Colorado Desert. Mexico in northern Baja California and northernmost Sonora.

SYNONYMY: Cereus Engelmannii Parry ex Engelm. *Echinocereus Engelmannii* Lemaire. Type locality: San Felipe, Colorado Desert, San Diego County, California.

5B. Var. acicularis
   L. Benson, **var. nov.** (p. 22)
   (Table 12)

DISTRIBUTION: Rocky, sandy or gravelly hillsides, washes, and valleys in the desert at 1,000 to 3,000 feet elevation. Arizona Desert and the edges of the Mojavean and Colorado deserts. Arizona from southern Mohave County near the Bill Williams River to the lower portions of Yavapai County and to Yuma, Maricopa, western Graham, Pinal, and western Pima counties. California in eastern Riverside County. Mexico in Baja California and Sonora.

Fig. 3.15. A yellow-spined hedgehog cactus, *Echinocereus Ledingii* in flower. (Photograph by Robert H. Peebles.)

A form of this variety occurring in Palm Canyon, Kofa Mountains, Yuma County, Arizona, forms colonies among rocks and on the flat tops of boulders and ledges where there is a little soil. These plants had up to 50 or more decumbent slender stems up to 2 feet long and about 2 inches in diameter.

5C. Var. *chrysocentrus*
   (Engelm.) Engelm. ex Rumpler
   (Table 12)

DISTRIBUTION: Gravel or sand of hills, low mountains, and washes in the desert at 3,000 to 5,000 or 7,200 feet elevation. Mojavean Desert. Arizona in Mohave County and northern Yuma County. California from the White Mountains in Inyo County to the higher desert mountains in Riverside County;

southwestern Nevada; Utah in the Great Salt Lake Desert (rare) and in Washington and southern Kane counties.

The long white deflexed lower central spine appears like a sword.

SYNONYMY: Cereus Engelmannii Parry var. *chrysocentrus* Engelm. & Bigelow. *Echinocereus Engelmannii* (Parry) Lemaire var. *chrysocentrus* Engelm. ex Rümpler. *Echinocereus chrysocentrus* Orcutt. *Echinocereus chrysocentrus* Thornber & Bonker. Type locality: lower part of [Bill] Williams River, Arizona. The lower deflexed central spine is white, as stated by Engelmann in a hand-written note accompanying the specimen. The other central spines are a dirty gray or blackened, as is commonly so with old once-yellow spines on the lower parts of cactus plants or in herbarium specimens. The epithet has been misapplied by several authors to var. *Nicholii* because of its yellow spines, the assumption being that var. *chysocentrus* must be yellow spined because of the name.

5D. Var. *variegatus*
   (Engelm.) Engelm. ex Rumpler
   (Table 12)

DISTRIBUTION: Rocky or gravelly hillsides in grasslands, woodlands, and deserts at 3,800 to 5,700 feet elevation. Great Plains Grassland, Juniper-Pinyon Woodland, and Navajoan Desert. Arizona from Cactus Pass, Mohave County, to the Navajo Bridge and Houserock Valley, Coconino County; rare in Gila County.

SYNONYMY: Cereus Engelmannii Parry var. *variegatus* Engelm. & Bigelow. *Echinocereus Engelmannii* (Parry) Lemaire var. *variegatus* Engelm. ex Rümpler. Type locality: head of Bill Williams River, 113½ degrees [West] longitude, Arizona.

5E. Var. *Nicholii*
   L. Benson
   (Table 12)

DISTRIBUTION: Gravel or sand of flats and hills in the desert at 1,000 to 3,000 feet elevation. Arizona and Colorado deserts. Arizona in Pima County from the Organ Pipe Cactus National Monument to the Silver Bell Mountains; reported by A. A. Nichol from the Mohawk Mountains in Yuma County. Mexico in adjacent Sonora.

SYNONYMY: Echinocereus Engelmannii (Parry) Lemaire var. *Nicholii* L. Benson. Type locality: Silver Bell Mountains, Pima County, Arizona.

TABLE 12. CHARACTERS OF THE VARIETIES OF ECHINOCEREUS ENGELMANNII

	A. Var. Engelmannii	B. Var. acicularis	C. Var. chrysocentrus	D. Var. variegatus	E. Var. Nicholii
Stems	5-15, not crowded, erect, 6 to 8 inches long, 2 inches in diameter.	Commonly 5-15 but up to 50 or more, usually erect, usually 6 to 8 inches long, 1½ to 2 inches in diameter.	3-10, erect, 5 to 8 or 13 inches long, 2 to 2½ inches in diameter.	3-6, erect, 3 to 6 inches long, 1½ to 2 inches in diameter.	Mostly 20-30, erect, not crowded, 12 to 24 inches long, 2 to 3 inches in diameter.
Spine color	Yellowish, pink, or gray, the lower central a little lighter.	Pinkish or yellowish, the lower central like the others.	Reddish to reddish-brown or yellow, dark or light, the lower central white.	Dark red to nearly black, but the radials nearly white, the lower central white.	All clear yellow or sometimes straw or nearly white, becoming black or gray in age.
Lower deflexed and flattened central spine	1¼ to 1¾ inches long, basally about 1/20 inch broad, stout, rigid, nearly straight.	1 to 1½ inches long, basally up to 1/24 inch broad, rather weak and flexible, straight.	1½ to 2¼ inches long, basally 1/20 to 1/16 inch broad, stiff and swordlike, straight or a little curved or twisted.	About 1½ inches long, basally up to 1/24 inch broad, stiff for its size, nearly straight.	2 to 2½ inches long, basally 1/20 to 1/16 inch broad, stiff, straight.
Other central spines	Straight, the largest nearly equalling the lower central, basally about 1/24 inch in diameter.	Straight, weak and flexible, up to 1 inch long, basally about 1/48 inch in diameter.	Straight, shorter than the lower central, usually dark-colored, but sometimes yellow, basally about 1/24 inch in diameter.	Straight, shorter than the lower central, dark-colored at least when young, about 1/48 inch in diameter.	Straight, a little shorter than the lower central, about 1/24 inch in diameter.
Flower diameter	2 to 2½ inches.	About 2½ inches.	2½ to 3 inches.	About 2 inches.	2 to 2½ inches.
Petaloid perianth segments	Magenta.	Purplish to magenta.	Purplish to magenta.	Purplish to magenta.	Lavender, much paler than in the other varieties.
Geographical distribution	Mostly Colorado Desert. Rare but widely distributed in Arizona. California from southern San Bernardino County to San Diego County and western Imperial County. Mexico in Baja California.	Mostly Arizona Desert. Arizona from near the Bill Williams River to Yuma, Yavapai, western Graham, and western Pima counties. California in eastern Riverside County.	Mojavean Desert. Arizona in Mohave County and northwestern most Yuma County. California from the White Mountains, Inyo County, to Riverside County and eastward to western Utah.	Grassland, woodland, and Navajoan Desert. Arizona from Mohave County to Coconino County and rare in Gila County.	Arizona Desert. Arizona in southeastern Yuma County and western Pima County. Mexico in adjacent Sonora.

Fig. 3.16. A hedgehog cactus, *Echinocereus Engelmannii.*
(Photographs by W. S. Phillips.)

Fig. 3.17. A hedgehog cactus, *Echinocereus Engelmannii*. This is the common hedgehog cactus of the Colorado Desert in California and Arizona. It is occasional in occurrence in the Arizona Desert, where, for the most part, it is replaced by var. *acicularis*. (Photographs: *above,* by A. A. Nichol; *below,* by A. R. Leding, U.S. Acclimatization Station, State College, New Mexico, SC 257; used by permission).

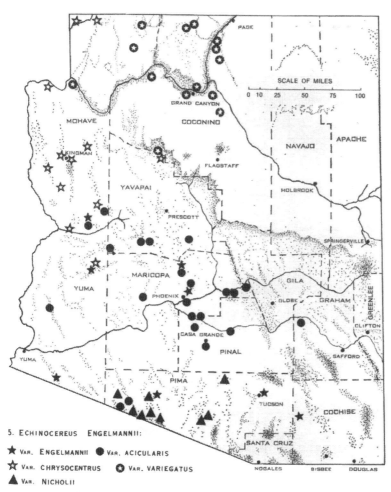

Fig. 3.18. The documented distribution of *Echinocereus Engelmannii*, according to its varieties.

5. ECHINOCEREUS ENGELMANNII:

★ VAR. ENGELMANNII ● VAR. ACICULARIS

☆ VAR. CHRYSOCENTRUS ⊕ VAR. VARIEGATUS

▲ VAR. NICHOLII

6. *Echinocereus pectinatus*
   (Scheidw.) Engelm.

Stem solitary or with two or three branches, the larger ones green, cylindroid, 4 to 12 inches long, 2½ to 4 inches in diameter; ribs 15 to 22; areoles elliptic to narrowly elliptic, 3/16 to 1/8 inch long; spines dense, obscuring the stem, pink to pale gray or straw-color or brown or some of them white; central spines 0 to 9 per areole, in 1 to 3 vertical series, at right angles to the stem, straight, the longer ones 1/24 to 5/16 inch long, basally 1/72 to 1/48 inch in diameter, needlelike, circular in cross section; radial spines 12 to 22 per areole, strongly pectinate to spreading irregularly, straight or curving downward in a low arc, the longer ones ¼ to ⅝ inch long, basally 1/96 to 1/48 inch in diameter, needlelike, nearly circular in cross section; flower 2½ to 5 inches in diameter; petaloid perianth parts magenta to light purple or lavender or yellow, sometimes magenta-and-yellow, the largest ones oblanceolate but with the apex rounded, 1¾ to 2½ inches long, up to ½ or ⅝ inch broad, apically rounded to acute, mucronate, dentate; fruit green or greenish-purple, with the clusters of spines deciduous, subglobose or ellipsoid, ¾ to 2½ inches long; ⅝ to 1¾ inches in diameter; seed 1/20 inch long.

6A. Var. *rigidissimus*
    (Engelm.) Engelm. ex Rumpler
    (Table 13)

DISTRIBUTION: Limestone hills chiefly in grasslands at 4,000 to 5,200 feet elevation.

Fig. 3.19. A hedgehog cactus, *Echinocereus Engelmannii* var. *acicularis*. This is the common variety in central Arizona. The spines are slender and yellowish or pink. (Photographs by Robert A. Darrow.)

Fig. 3.20. Arizona rainbow cactus, *Echinocereus pectinatus* var. *rigidissimus*. The flowers are pink to lavender. There are no central spines. (Photograph by R. B. Streets.)

Desert Grassland and the lower edge of the Southwestern Oak Woodland. Southern Arizona near Phoenix and Tempe in Maricopa County and in Pima, Santa Cruz, and Cochise counties. New Mexico in Grant and Hidalgo counties. Mexico in northern Sonora and Chihuahua.

Specimens from Eddy County, New Mexico, and from near Alpine, Brewster County, Texas, approach this variety.

SYNONYMY: Cereus pectinatus (Sheidw.) Engelm. var. *rigidissimus* Engelm. *Echinocereus pectinatus* (Scheidw.) Engelm. var. *rigidissimus* Engelm. ex. Rümpler. *Echinocereus dasyacanthus* Engelm. var. *rigidissimus* W. T. Marshall. Type locality: Sierra del Pajarito, Sonora.

6B. Var. *pectinatus*
(Table 13)

DISTRIBUTION: Limestone hills and flats in the desert and in grasslands at 3,500 to 4,500 elevation. Chihuahuan Desert and Desert Grassland. Southeastern corner of Arizona (in Cochise County from the Mule Mountains eastward). Texas, rare near the Rio Grande as far east as Eagle Pass. Mexico southward as far as San Luis Potosí.

SYNONYMY: Echinocactus pectinatus Scheidw. *Echinopsis pectinata* Fennel. *Echinocereus pectinatus* Engelm. *Cereus pectinatus* Engelm. Type locality: "Habitat propre l'ida del Pennasco in locis temperatis." *Cereus ctenoides* Engelm. *Echinocereus*

145

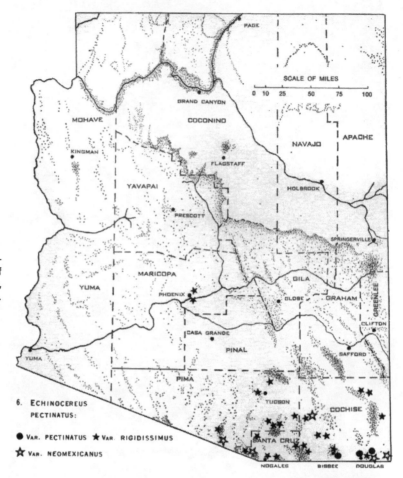

Fig. 3.21. The documented distribution of *Echinocereus pectinatus*, according to its varieties.

6. ECHINOCEREUS PECTINATUS:

● VAR. PECTINATUS ★ VAR. RIGIDISSIMUS
☆ VAR. NEOMEXICANUS

TABLE 13. CHARACTERS OF THE VARIETIES OF ECHINOCEREUS PECTINATUS

	A. Var. **rigidissimus**	B. Var. **pectinatus**	C. Var. **neomexicanus**
Central spines	None.	3-5 in 1 or sometimes 2 vertical series, 1/24 to 1/8 inch long.	3 or 7-9 in 2 or 3 vertical series, 1/8 to 5/16 inch long.
Radial spines	18-22, white and red in alternating horizontal color bands on the stem, obviously pectinate, the longer ones 5/16 to 3/8 inch long, stout, 1/48 inch in diameter.	12-16, pink or pink-and-gray, spreading somewhat irregularly, the longer ones 1/4 to 5/16 inch long, rather stout, about 1/96 to 1/72 inch in diameter.	18-22, pale brown or pink, spreading rather irregularly, the longer ones 3/8 to 1/2 inch long, of moderate thickness, about 1/96 to 1/72 inch in diameter.
Flowers	Magenta, 2½ to 3½ inches in diameter.	Magenta, 2½ to 3½ inches in diameter.	Yellow, 3 to 5 inches in diameter.
Geographical distribution	Southern Arizona in Pima, Santa Cruz, and Cochise counties. Mexico in northern Sonora.	Arizona (southeasternmost corner). Texas near the Rio Grande as far east as Maverick County. Mexico, southward as far as San Luis Potosi.	Arizona near Benson, Cochise County. New Mexico from Valencia County to Chaves County and southward; Texas from El Paso County to Culberson and Brewster counties. Mexico in northern Chihuahua and perhaps northwestern Coahuila.

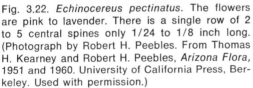

Fig. 3.22. *Echinocereus pectinatus.* The flowers are pink to lavender. There is a single row of 2 to 5 central spines only 1/24 to 1/8 inch long. (Photograph by Robert H. Peebles. From Thomas H. Kearney and Robert H. Peebles, *Arizona Flora,* 1951 and 1960. University of California Press, Berkeley. Used with permission.)

Fig. 3.23. Yellow-flowered hedgehog cactus, *Echinocereus pectinatus* var. *neomexicanus.* Petaloid perianth parts are yellow. Usually 7 to 9 central spines in 2 or 3 rows; spines range from 1/8 to 5/16 inch long. (Photograph by A. R. Leding. U.S. Acclimatization Station, State College, New Mexico. SC 155. Used with permission.)

ctenoides Engelm. ex Rümpler. *Echinocereus dasya-canthus* Engelm. var. *ctenoides* Backeberg. Type collection: Santa Rosa, Coahuila.

6C. Var. *neomexicanus*
   (Coulter) L. Benson
   (Table 13)

DISTRIBUTION: Limestone hills and flats in the desert and grasslands at 4,000 to 5,000 feet elevation. Chihuahuan Desert and Desert Grassland. Arizona near Benson, Cochise County. Although the reports are undocumented, this variety has been reported to occur also (according to J. J. Thornber) in the southern part of the Baboquivari Mountains, Pima County, and (according to A. A. Nichol)in the Perilla and Guadalupe mountains, Cochise County, as well as (according to A. A. Nichol) in perhaps modified form in the Oro Blanco Mountains west of Nogales, Santa Cruz County, and at Altar and Pitiquito in adjacent Sonora. Southeastern New Mexico; Texas from El Paso County to Culberson and Brewster counties. Mexico in northern Chihuahua and perhaps in northwestern Coahuila.

The varieties of *Echinocereus pectinatus* have been divided arbitrarily into species or even groups of species on the basis of flower color. Authors have assumed flower color to be a "fundamental character" always separating the "species." Like all other characters, color may occur in combination with any other character or characters, being inherited in the same way as other genetic features. Always overemphasis of a single character results in an artificial system of classification. Intermediate coloring involving both yellow and pink-to-light-purple pigments in the petaloid perianth parts occurs in a number of populations observed in the field and in specimens.

The most striking other difference between vars. *neomexicanus* and *pectinatus* is the greater number and size of central spines and the disorderly placement of radial spines. This produces superficially different plants. The central spines are in two or three irregularly crowded rows in var. *neomexicanus,* and, in keeping with their small number, usually they are in a single row in var. *pectinatus.* When in var. *pectinatus* the number of central spines is greater than usual they are crowded out into two irregular rows. Radial spines of var. *pectinatus* are

intermediate between the disarray of var. *neomexicanus* and the primly pectinate arrangement of var. *rigidissimus,* which is without central spines.

SYNONYMY: Echinocereus dasyacanthus Engelm. *Cereus dasyacanthus* Engelm. Type locality: El Paso, Texas. *Cereus dasyacanthus* Engelm. var. *neomexicanus* Coulter. *Echinocereus pectinatus* (Scheidw.) Engelm. var. *neomexicanus* L. Benson, not *Echinocereus neomexicanus* Standley. As explained by the writer, "The necessity for use of this name [epithet] in varietal status is unfortunate, for the little-known epithet *neomexicanus* was proposed later in specific rank for a plant of the *Echinocereus triglochidiatus* complex." Type locality: "Southeastern New Mexico." *Echinocereus spinosissimus* Walton. Type locality: El Paso, Texas. *Echinocereus rubescens* Dams. *Echinocereus papillosus* Linke var. *rubescens* [Dams?], *pro syn.* Named from cultivated material; no specimen or origin given.

4. MAMMILLARIA

Stems simple or branching, the mature ones ovoid to cylindroid, globose, or turbinate, 1 to 4 or 12 inches long, mostly 1 to 3 or up to 8 inches in diameter; tubercles separate. Leaves not discernible on the mature tubercle. Spines smooth, with a broad range of colors; central spines 0 to several or sometimes not differentiated, straight, curved, or hooked, mostly ¼ to ¾ or 1 inch long, slender, basally usually 1/240 to 1/72 inch in diameter, needlelike, broadly elliptic in cross section; radial spines smaller and of a lighter color, 10 to 80 per areole, not hooked, needlelike. Flowers and fruits developed on the old growth and therefore located below the apex of the stem or branch, between the tubercles and not obviously connected with them or with the spine-bearing areoles upon them, sometimes in minor spine-bearing areoles upon them, sometimes in minor spine-bearing areoles between tubercles. Flower usually ¼ to 1 or rarely 2 inches in diameter; floral tube above its junction with the ovary funnelform or obconic, usually green tinged with the perianth color. Fruit fleshy, without surface appendages, globular to elongate, usually ¼ to 1 inch long, ¼ to ¾ inch in diameter, with the floral tube deciduous, indehiscent. Seeds black to brown, rugose-reticulate, reticulate-pitted, smooth and shiny or tuberculate, longer than broad (length being hilum to opposite side), 1/24 to 1/12 inch long.

DISTRIBUTION: Perhaps 100 or more species (far greater numbers having received names) from California to southern Nevada, southwestern Utah, New Mexico, western Oklahoma, and Texas and thence to Mexico and Central America; rare in the Caribbean and Venezuela. Fourteen species in the United States. Eight in Arizona.

LIST OF SPECIES AND VARIETIES

KEY TO THE SPECIES

1. Juice of the tubercles milky; stem top-shaped or depressed-globose, the apex flattened or shallowly convex, often barely protruding above the surface of the ground; tubercles conical or pyramidal, not obscured by spines; spines straight or slightly curving, *none* recurving like fishhooks.
   1. *Mammillaria gummifera,* page 149
1. Juice of the tubercles not milky; stem cylindroid or ovoid, the apex *not* markedly flattened or shallowly convex, rising well above the sur-

face of the ground (except in *Mammillaria Mainiae,* which has hooked, yellow central spines); tubercles conical to cylindroidal, obscured by spines.

2. Spines (one or more of the centrals but not the radials) hooked.

   3. Base of the seed not corky, the aril, if any (a structure about the hilum or attachment-scar) *not* markedly thickened or corky; hooked central spine 1 or sometimes 2 (or 3 in *Mammillaria Wrightii*); radial spines 8-35.

      4. Central spines yellow except at the tips, basally 1/36 to 1/48 inch in diameter, the hooks nearly all turned counterclockwise; stigmas red; sepals with very short hairs on the margins, the hairs much less than 1/32 inch long.

         2. *Mammillaria Mainiae,* page 152

      4. Central spines *not* wholly or partly yellow, the hooks not turned consistently in any direction.

        5. Apical hook of the central spine 1/12 to ⅛ inch across (except in immature plants); stem firm, ending abruptly at the base instead of tapering into the root, often branching, the clusters of stems thus formed from a single plant; stigmas green; flowers ¾ to 1⅛ inches in diameter.

           3. *Mammillaria microcarpa,* page 152

        5. Apical hook of the central spine 1/16 inch across; stem flabby, rarely branching, the clusters of stems, if any, usually of separate individuals.

          6. Only 1 of the 1 to 3 central spines hooked and dark reddish-brown, the others, if any, shorter, pale, and not hooked; radial spines 15 to 35 per areole, 1/4 to 7/16 inch long, 1/120 inch in diameter; seed with a smooth shiny surface, this more prominent than the pits in the network or meshwork pattern of the surface.

          7. Central spine 1, hooked (sometimes in a few areoles accompanied by one shorter, paler, straight accessory upper central); flower ⅝ to ¾ inch in diameter; petaloid perianth parts up to ⅜ or ½ inch long; seed black, 1/24 inch long; radial spines pale tan.

          8. Stem mostly a slender cylinder but tapering gradually into the base, usually 2 to 4 but up to 10½ inches long, ½ to 1 or in long stems up to 2 inches in diameter; stigmas 7, bright red, 1/8 to 1/6 inch long, petaloid perianth parts lavender; anthers rectangular, 1/32 inch long; radial spines brown-tipped; sepals finely ciliate.

           4. *Mammillaria Thornberi,* page 155

        8. Stem ovoid, 3 to 4 inches long, 2 to 3 inches in diameter; stigmas 4 to 5, green, 1/24 inch long and relatively broad; petaloid perianth parts green or green tinged with pink, anthers nearly square, 1/72 inch long; sepals strongly fimbriate.

           5. *Mammillaria orestera,* page 155

        7. Central spines 3, the chief one hooked and longer than the two accessory short, pale, straight upper ones; flower ¾ to 1¾ inches in diameter; petaloid perianth parts ½ to 1 inch long, white to pale pink or lavender; seed dark brown, 1/16 inch long; radial spines white or pale tan; stigmas 1/16 or 1/4 to 5/16 inch long.

           6. *Mammillaria Grahamii,* page 159

      6. Central spines all (3 or rarely only 2) hooked, dark reddish-brown; radial spines about 10-12, about 1/72 inch in diameter, light tan; seed with the pits in the surface more prominent than the reticulate pattern, 1/16 inch long, brown.

         7. *Mammillaria Wrightii,* page 161

   3. Base of the seed corky, the corky portion (the enlarged aril or outgrowth next to the attachment-scar or hilum) almost as large as the seed-body; hooked central spines 1-4; radial spines 30-46 or 60; deserts near the Colorado River.

      8. *Mammillaria tetrancistra,* page 161

2. Spines not hooked; southeastern Arizona.

   3'. Spines 20 to 30 per areole, in one series of radial and one of central spines, these strongly differentiated, the centrals 3; stems cylindroid to subglobose, mostly 3 to 5 inches long, 2 to 3 inches in diameter; flower campanulate to funnelform; petaloid perianth parts pink to reddish-purple.

        6b. *Mammillaria Grahamii* var. *Oliviae,* page 161

   3'. Spines 40 or commonly 50 to 80 per areole, in several undifferentiated series; stem globose, the diameter not more than 1½ inches; flowers rotate (saucer-shaped), ⅜ to ½ inch in diameter; petaloid perianth parts white with red midstripes.

        9. *Mammillaria lasiacantha,* page 163

1. *Mammillaria gummifera*
Engelm.

Stems low-growing, usually solitary, the larger ones top-shaped to subglobose, the portion of each above ground level flat or depressed or shallowly convex, 3 to 4 or 6 inches in diameter; tubercles subconical to subpyramidal; spines dense but not obscuring

Fig. 4.1. *Mammillaria gummifera* var. *MacDoug-alii* in flower. The plant has milky juice. (Photograph by A. A. Nichol.)

the stem; central spines brown or reddish-brown or tan, 0, 1, or 2 per areole, standing at right angles to the surface, straight, the longer ones 1/8 to 3/8 inch long, basally

1/120 to 1/48 inch in diameter, nearly circular in cross section; radial spines tan to brown or nearly white, 6 to 22 per areole, spreading parallel to the stem, straight, the longer ones 1/4 to 9/16 inc hlong, basally 1/120 to 1/48 inch, nearly circular in cross section; flower ½ to 1¼ inches in diameter; petaloid perianth parts pink, white, or cream or in mixtures of these colors, the largest linear-lanceolate, 3/8 to 3/4 inch long, about 1/16 to 1/8 or 3/16 inch broad, acute to acuminate or mucronate, entire to denticulate; fruit red, fleshy, narrowly obovoid, markedly enlarged upward, ½ to 1½ inches long, ¼ to ⅜ inch in diameter; seeds brown, rugose-reticulate, 1/24 inch long.

1A. Var. **applanata**
      (Engelm.) L. Benson, **comb. nov.** (p. 22)
      (Table 14)

DISTRIBUTION: Rocky or gravelly limestone soils in the desert and in grasslands in Arizona at 4,000 to 4,500 feet elevation. Chihuahuan Desert and adjacent grasslands. Southeasternmost corner of Arizona in Cochise County. Southern New Mexico; Texas from El Paso County to San Patricio County and southward to the Rio Grande. Northern Mexico through an undetermined area.

TABLE 14. CHARACTERS OF THE VARIETIES OF MAMMILLARIA GUMMIFERA

	A. Var. **applanata**	B. Ver. **meiacantha**	C. Var. **MacDougalii**
Stem apex	Flattened to concave.	Hemispheroidal.	Flattened or concave to hemispheroidal.
Tubercles	Subconical.	Subpyramidal.	Subconical.
Central spine(s)	1, ¼ to 5/16 inch long.	0 or 1, ⅛ to ¼ inch long.	1 or 2, 5/16 to ⅜ inch long.
Radial spines	10 or 14-22, 5/16 to ⅜ inch long, 1/120 inch in diameter.	6-9, 5/16 to 7/16 inch long, 1/48 inch in diameter.	10-12, 5/16 to 9/16 inch long, 1/72 inch in diameter.
Flower color and size	Pink to nearly white or cream, up to ¾ inch in diameter.	Pink or pink-and-white, up to 1 or 1¼ inches in diameter.	Cream, up to 1¼ inches in diameter.
Margins of the sepaloid perianth parts	Entire.	Entire or minutely denticulate.	More or less fimbriate (fringed).
Geographical distribution	Southeasternmost Arizona. Southwestern New Mexico; most of Texas. Mexico.	Arizona in Texas Canyon, Dragoon Mountains, Cochise County. Southcentral New Mexico; Texas west of the Pecos. Mexico (upland).	Southeastern Arizona. Mexico in adjacent Sonora.

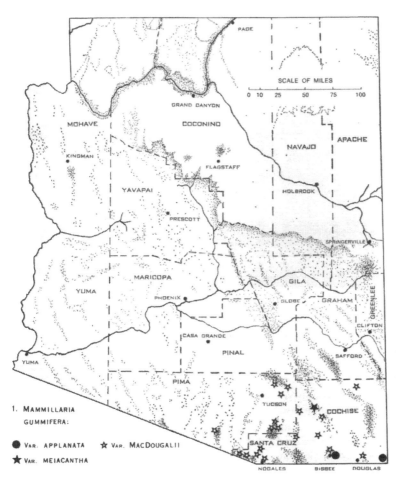

Fig. 4.2. The documented distribution of *Mammillaria gummifera*, according to its varieties.

SYNONYMY: Mammillaria applanata Engelm., Jan. 13, 1848. *Mammillaria Heyderi* Mühlenpfordt var. *applanata* Engelm. *Neomammillaria applanata* Britton & Rose. Type locality: Perdenales River, Texas. *Mammillaria Heyderi* Mühlenpfordt, Jan. 15, 1848. *Cactus Heyderi* Kuntze. *Neomammillaria Heyderi* Britton & Rose. Type locality: No locality or specimen mentioned. According to Roemer, "Am Llanoflusse," Texas. *Mammillaria declivis* Dietr. Type locality: "Habitat in Texas." *Mammillaria texensis* Labouret. *Cactus texensis* Kuntze. Type locality: Texas.

1B. Var. **meiacantha**
(Engelm.) L. Benson, **comb. nov.** (p. 22)
(Table 14)

DISTRIBUTION: Gravelly or rocky soils, usually of limestone origin, in the desert and grasslands at 4,000 to 5,300 feet elevation. Chihuahuan Desert, Great Plains Grassland, and Desert Grassland. Arizona in Texas Canyon, Dragoon Mountains, Cochise County.

New Mexico; Texas from El Paso County to Jeff Davis and Brewster counties. Mexico as far south as Zacatecas.

SYNONYMY: Mammillaria meiacantha Engelm. *Cactus meiacanthus* Kuntze. *Neomammillaria meiacantha* Britton & Rose. *Mammillaria melanocentra* Poselger var. *meiacantha* Craig. Type locality: "Cedar plains east of the Pecos," New Mexico.

1C. Var. **MacDougalii**
(Rose) L. Benson, **comb. nov.** (p. 22)
(Table 14)

DISTRIBUTION: Hillsides, mountain valleys, and on plains in grassland at 3,600 to 5,000 feet elevation. Desert Grassland. Arizona along the southern edge of Pinal County and from Pima County to Santa Cruz and Cochise counties. Mexico in adjacent Sonora.

SYNONYMY: Mammillaria MacDougalii Rose. *Neomammillaria MacDougalii* Britton and Rose.

Fig. 4.3. *Mammillaria Mainiae* in flower. (Photograph by A. A. Nichol.)

Mammillaria Heyderi Mühlenpfordt var. *MacDougalii* L. Benson. Type locality: Santa Catalina Mountains, Pima County, Arizona.

2. *Mammillaria Mainiae*
    K. Brandegee

Stem gray-green or blue-green, usually solitary, hemispheroid or ovoid to subglobose, 2½ to 5 inches long, 2 to 3 inches in diameter; tubercles mammaeform-cylindroid; spines rather dense, partly obscuring the stem; central spine yellow with a brown tip, solitary in the areole (reportedly sometimes with 2 additional straight centrals), hooked, the central spines turned counterclockwise in the areoles around the stem, ½ to ¾ inch long, basally 1/36 to 1/48 inch in diameter, nearly circular in cross section; radial spines yellow with brown tips, 10 to 15 per areole, spreading parallel to the stem surface, straight, the longer ones 1/4 to 3/8 inch long, basally 1/120 inch in diameter; flower about 3/4 to 1 inch in diameter; petaloid perianth parts red in the middles and white on the margins, lanceolate, spreading; fruit red, fleshy at maturity, obovoid, 1/4 to 3/8 inch long, 1/8 to 3/16 inch in diameter; seed black, *not* shiny, the surface more conspicuous than the pits in the meshwork, 1/24 inch long.

DISTRIBUTION: Gravel or coarse sandy soil of hills, washes, and alluvial fans in the

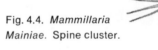

Fig. 4.4. *Mammillaria Mainiae.* Spine cluster.

desert or in grassland at 2,000 to 4,000 feet elevation. Sonoran Deserts and Desert Grassland. Arizona near the Mexican Boundary and on the Papago Indian Reservation, Pima County, and near Nogales, Santa Cruz County. Mexico in Sonora and Sinaloa.

SYNONYMY: Mammillaria Mainiae K. Brandegee. *Chilita Mainiae* Orcutt. *Ebnerella Mainiae* F. Buxbaum. Type locality: south of Nogales, Sonora.

3. *Mammillaria microcarpa*
    Engelm.

Stems solitary at first but often branching as the plant becomes older, cylindroid, the base at maturity truncate rather than tapering as in *Mammillaria Thornberi,* 3 to 5 or 6 inches long, 1½ to 2 inches in diameter; spines rather dense, partly obscuring the stem; central spine dark red or sometimes black-purple, the areole with 1 principal hooked central and sometimes another

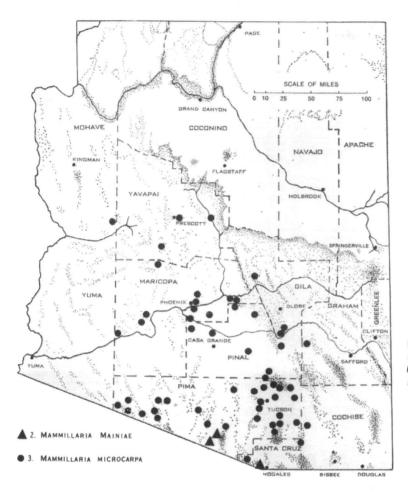

Fig. 4.5. The documented distribution of *Mammillaria Mainiae* and of *Mammilaria microcarpa*.

2. MAMMILLARIA MAINIAE

3. MAMMILLARIA MICROCARPA

straight, short, light-colored central, these spines spreading nearly at right angles to the stem, curving or essentially straight except for the hook (this 1/12 to 1/8 inch across), usually 1/2 to 5/8 inch long, basally 1/32 inch in diameter; radial spines light tan to red, 18 to 28 per areole, spreading parallel to the stem, straight, mostly ¼ to ½ inch long, basally 1/120 to 1/96 inch in diameter; flower ¾ to 1⅛ inches in diameter; petaloid perianth parts lavender, the largest ones oblanceolate, 1/2 to 5/8 inch long, 1/8 to 3/16 inch broad, short-acuminate, entire; fruit red, fleshy, cylindroid to club-shaped, tapering gradually upward, usually ½ to 1⅛ inches long, up to 3/16 inch in diameter; seeds black, the shiny surface more prominent than the pits that are in the meshwork, 1/32 to 1/24 inch long.

DISTRIBUTION: Sand and gravel of canyons, washes, alluvial fans, and plains in the desert or grassland at 1,000 to 3,500 or 5,000 feet elevation. Arizona Desert. Arizona from the southeastern corner of Mohave County to western Graham and Pima counties. Mexico at least near the borders of Arizona.

SYNONYMY: Mammillaria microcarpa Engelmann. *Neomammillaria microcarpa* Britton & Rose. *Ebnerella microcarpa* F. Buxbaum. *Chilita microcarpa* F. Buxbaum. Type locality: near Christmas, Gila River, Arizona. *Mammillaria Grahamii* Engelm. var. *arizonica* Quehl. Horticultural plant of unknown origin. *Neomammillaria Milleri* Britton & Rose. *Mammillaria Milleri* Bödeker. *Mammillaria microcarpa* Engelm. var. *Milleri* W. T. Marshall. Type locality: near Phoenix, Arizona. *Mammillaria microcarpa* Engelm. var. *auricarpa* W. T. Marshall. Type locality: near Pinnacle Peak, Maricopa County, Arizona.

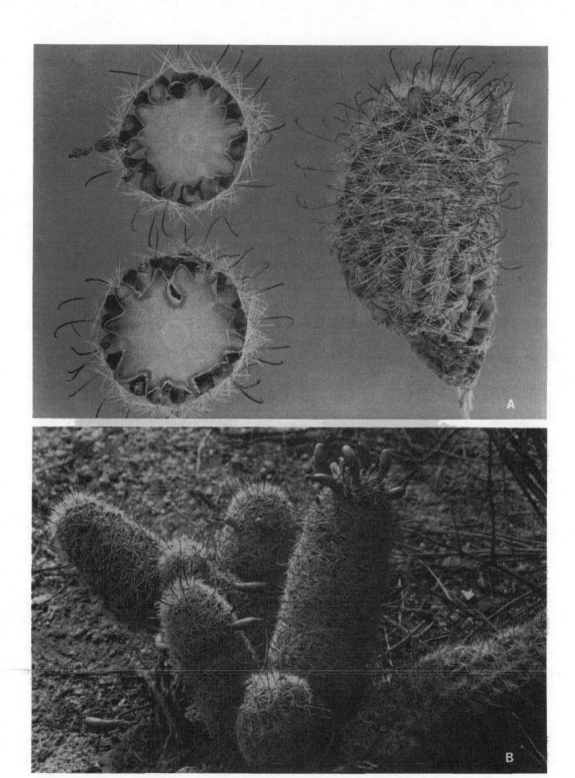

Fig. 4.6. *Mammillaria microcarpa. A,* plant with fruits of the elongate, red (mature) type and sections showing the barely immature globose, green fruits among the tubercles. These elongate rapidly and turn red at maturity. *B,* plant with mature red elongate fruits. (Photographs by Robert H. Peebles.)

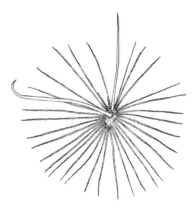

Fig. 4.7. *Mammillaria microcarpa.* Spine cluster with one hooked (lower) central spine and with a smaller one above it in the areoles.

4. *Mammillaria Thornberi*
Orcutt

Stem solitary, but the individuals closely associated in clumps of up to 100; stem cylindroid or narrowly cylindroid, usually 2 to 4 but sometimes 10½ inches long, ½ to ¾ or in very long stems up to 2 inches in diameter, tapering gradually into the root; spines of moderate density, partly obscuring the stem; central spine reddish-brown, darkened gradually apically, 1 per areole, spreading at right angles to the stem, apically hooked, with the hook 1/16 inch across, the longer spines 3/8 to 1/2 inch long, basally 1/64 inch in diameter; radial spines basally straw color, the tips reddish-brown, 15 to 20 per areole, spreading parallel to the stem, the longer ones ¼ to ⅜ inch long, basally about 1/120 inch in diameter; flower ⅝ to ¾ inch in diameter; petaloid perianth parts lavender, lanceolate, 3/8 to 1/2 inch long, 1/8 to 3/16 inch broad, acute, entire; fruit red, fleshy at maturity, 3/8 to 1/2 or 5/8 inch long, 3/16 to 5/16 inch in diameter; seed black, the shiny surface more prominent than the pits in the meshwork, 1/24 inch long.

DISTRIBUTION: Sandy or fine soils under shrubs in flats and washes in the desert at 800 to 2,400 feet elevation. Plants tolerant of alkaline conditions. Arizona Desert. Arizona in Pinal and (mostly) Pima counties; largely on the Papago Indian Reservation. Mexico in northern Sonora.

Commonly there are many stems growing together. These are the unbranched stems of individuals crowded against each other. The plant is not well known, but the few collections and study in the field have not revealed branching stems. This point was not noticed by Britton and Rose (Cactaceae 4:162. 1923) when they sought to apply the name *Mammillaria fasciculata* to this species. Neither was the geographical and ecological separation of *Mammillaria Thornberi* from the plants upon which *Mammillaria fasciculata* was based, *i.e., Echinocereus fasciculatus,* p. 132.

SYNONYMY: Mammillaria Thornberi Orcutt. *Chilita Thornberi* Orcutt. Neotype locality: west of the Silver Bell Mountains, Pima County, Arizona.

5. **Mammillaria orestera**
L. Benson, **sp. nov.** (p. 22)

Stems solitary, or rarely branching, orbicular to ovoid, the larger ones 3 to 4 inches long, 2 to 3 inches in diameter; spines dense, more or less obscuring the stem; central spines reddish-brown, usually 1 per areole or in addition occasionally with one weaker, short, pale, straight central present near the upper margin of the areole, the principal central projecting and somewhat curving, apically hooked (the hook 1/16 inch across), 3/8 to 1/2 inch long, basally 1/96 inch in diameter; radial spines pale tan, 15 to 24 per areole; spreading parallel to the stem, straight, up to 1/3 to 7/16 inch long, basally 1/120 inch in diameter; flower ⅝ to ¾ inch in diameter; petaloid perianth parts green or green tinged with pink, the largest lanceolate, up to ⅜ inch long, 1/12 inch broad, acute, entire or faintly toothed; fruit red, fleshy at maturity, subglobose or elongate, ⅜ to ⅝ or ¾ inch long, about 3/16 to 1/4 inch in diameter; seed black, smooth and shiny, the surface more prominent than the pits, 1/24 inch long.

DISTRIBUTION: Sandy granitic soils on mountainsides in oak woodland and along the edges of deserts and forests at 3,000 to 6,500 feet elevation. Southwestern Oak Woodland and Chaparral. Arizona in the Santa Catalina Mountains, Pima County, and in the Graham Mountains, Graham County.

Type locality: Santa Catalina Mountains, Pima County, Arizona.

Fig. 4.8. *Mammillaria. Above, Mammillaria Grahamii,* with the hook of the slender central spine about 1/16 inch across. *Below, Mammillaria microcarpa,* with the hook of the stout central spine 1/12 to 1/8 inch across.

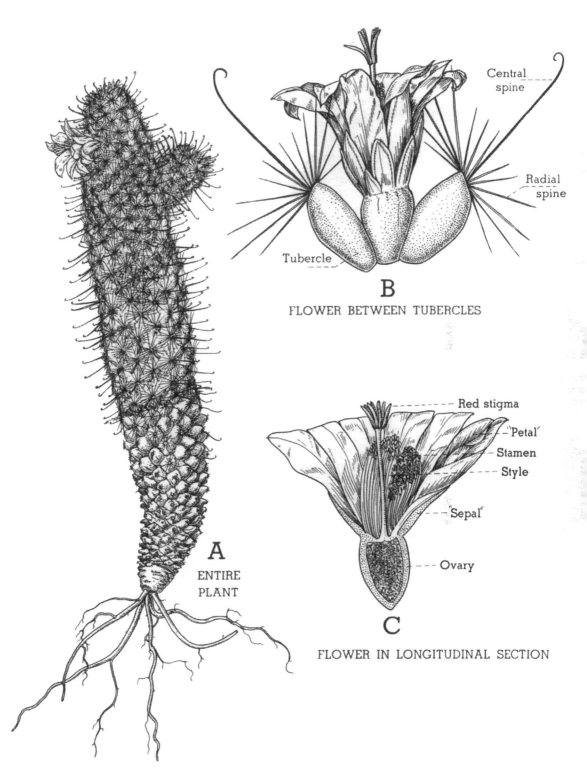

Central
spine

Radial
spine

Tubercle

B

FLOWER BETWEEN TUBERCLES

Red stigma

"Petal"

Stamen

Style

"Sepal"

Ovary

C

FLOWER IN LONGITUDINAL SECTION

A
ENTIRE
PLANT

Fig. 4.9. *Mammillaria Thornberi,* a species occurring northwest and west of Tucson on the Papago Indian Reservation.

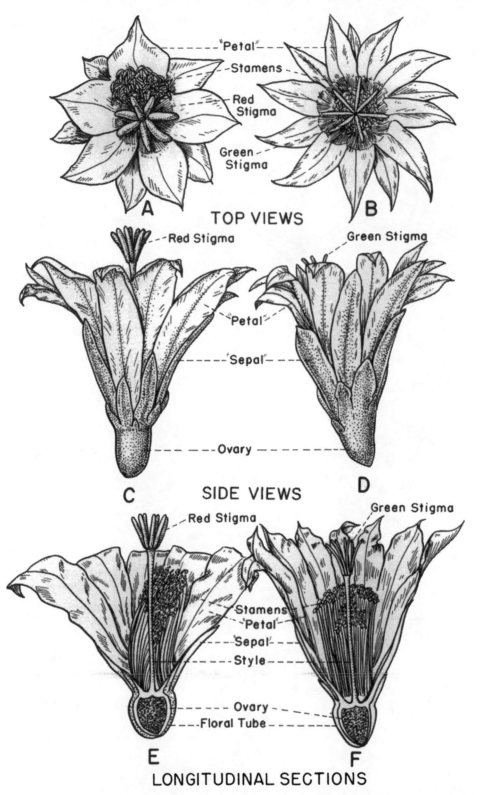

"Petal"

Stamens

Red
Stigma

Green
Stigma

A **B**

TOP VIEWS

Red Stigma Green Stigma

Petal

"Sepal"

Ovary

C **D**

SIDE VIEWS

Red Stigma Green Stigma

Stamens
"Petal"
"Sepal"
Style

Ovary
Floral Tube

E **F**

LONGITUDINAL SECTIONS

Fig. 4.10. Comparison of the flowers (A, C, E) of *Mam-millaria Thornberi* and (B, D, F) *Mammillaria microcarpa.*

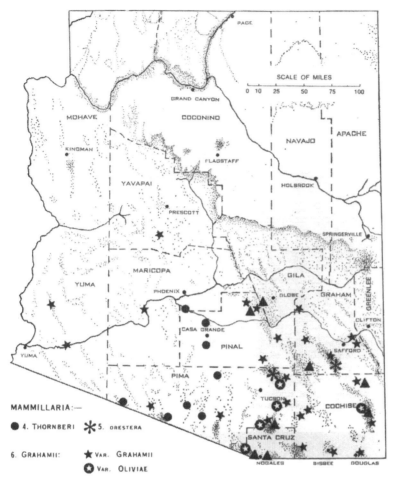

MAMMILLARIA:—

● 4. THORNBERI   ✳ 5. ORESTERA

6. GRAHAMII: ★ VAR. GRAHAMII
              ✪ VAR. OLIVIAE

Fig. 4.11. The documented distribution of *Mammillaria Thornberi*, *Mammillaria orestera*, and *Mammillaria Grahamii*, according to its varieties.

6. *Mammillaria Grahamii*
Engelm.

Stems solitary, green, ovoid to spheroid, 2 to 3 or up to 4 inches long, 1¾ to 2½ or up to 4½ inches in diameter; spines rather dense, partly obscuring the stem; central spines dark reddish-brown or red to nearly black, 2 or commonly 3 to 4 per areole, spreading, in var. *Grahamii* the principal one (central in the areole) recurved at the apex and curving somewhat below, the hook about 1/16 inch or less straight across from the point, this central ½ to 1 inch long, basally 1/120 inch in diameter, the others (upper marginal centrals) much shorter, less colored, and straight, in var. *Oliviae* all or nearly all the centrals white and *not* hooked and only ⅙ to ⅓ inch long; radial spines white, 20 to 35 per areole, spreading closely at right angles to the stem, straight, up to ¼ or ⅓ inch long, basally

1/240 to 1/120 inch in diameter; flower ¾ to 1¼ inches in diameter; petaloid perianth parts about 20, white or pale pink, the largest ones narrowly lanceolate, ½ inch long, about 1/12 inch broad, sharply acute, entire to ciliate-fimbriate; beshy, subglobose or barrel-shaped, ½ to 1 inch in diameter; seeds black, smooth and shining, the small pits less prominent than the shiny surface of the meshwork that lies between the pits, 1/20 to 1/16 inch long.

6A. Var. *Grahamii*

Central spines 1 to 3, dark reddish-brown or red to nearly black, the longest hooked and slightly curving, ½ to 1 inch long.

DISTRIBUTION: Hills and washes in grassland at 3,000 to 5,000 feet elevation. Primarily desert grassland from Yavapai County to southern and eastern Pima County to and

Fig. 4.12. *Mammillaria. Above, Mammillaria Grahamii* var. *Oliviae* in flower. The plant has a few hooked central spines, and it is transitional to var. *Grahamii. Below, Mammillaria Wrightii* in flower.

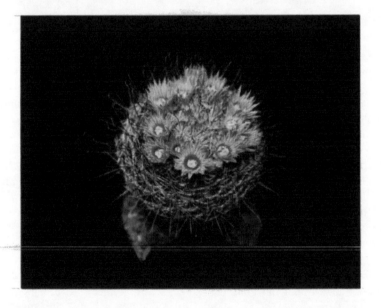

including Cochise County. Southwestern New Mexico; Texas in El Paso County and (rare) in Presidio County. Where their ranges overlap var. *Grahamii* intergrades with *Mammillaria microcarpa*.

SYNONYMY: *Mammillaria Grahamii* Engelm. *Cactus Grahamii* Kuntze. *Coryphantha Grahamii* Rydb. *Chilita Grahamii* Orcutt. Type locality: Franklin Mountains, El Paso, Texas.

6B. Var. **Oliviae**
(Orcutt) L. Benson, **comb. nov.** (p. 22)

Central spines white or the lower one sometimes dark-tipped, 3, not straight and usually none hooked or curved, up to ⅙ to ⅓ inch long, the two upper longer than the lower (in the center of the areole).

DISTRIBUTION: Rocky or gravelly soils in the desert and the lower edge of the grassland at about 3,000 feet elevation. Arizona Desert and Desert Grassland. Arizona in eastern Pima County and (rare) in Santa Cruz and Cochise counties. Mexico in northern Sonora.

A few collections have some hooked central spines. Because of these the plant was supposed to be a form of *Mammillaria microcarpa*. However, the occasionally occurring hooked centrals in some areoles are of the type in *Mammillaria Grahamii*. Both the lectotype and the duplicate mentioned include a few spines of the *Mammillaria Grahamii* type. A collection from an area adjacent to the type locality near the Colossal Cave includes material of both varieties for comparison. The flowers are of the *Mammillaria Grahamii* type.

SYNONYMY: *Mammillaria Oliviae* Orcutt. *Neomammillaria Oliviae* Britton & Rose. *Chilita Oliviae* Orcutt. *Ebnerella Oliviae* F. Buxbaum. Type locality: near Vail, Pima County, Arizona.

7. *Mammillaria Wrightii*
Engelm.

Stem green, solitary, globose or depressed-globose or top-shaped, 1½ to 4 inches long, 1½ to 3 inches in diameter; spines rather dense, partly obscuring the stem; central spines dark reddish-brown, 3 (or rarely 2) per areole, spreading, all somewhat curving and hooked, equal, about ½ inch long, basally 1/120 to 1/72 inch in diameter; radial spines at first tan, later becoming gray, about 10 to 20 per areole, parallel to the stem, straight, the longer ones 1/2 inch long, basally 1/72 inch in diameter; flower 1½ to 1¾ inches in diameter; petaloid perianth parts about 20

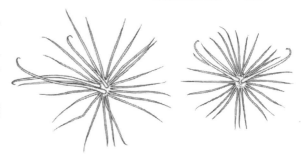

Fig. 4.13. *Mammillaria Wrightii*. Spine clusters. In mature plants there are usually 3 hooked central spines, in young plants, 1 or 2.

("12"), reddish-purple, linear-lanceolate, 1 to 1¼ inches long, ⅛ inch broad, acute, entire; fruit red, ovoid or nearly orbicular, up to 1 inch in diameter; seeds brown, with the pits in the meshwork more prominent than the surface between them, 1/16 inch long.

Similar to *Mammillaria Grahamii*, but differing as follows: Stem usually globose to depressed-globose or turbinate, 1½ to 3 inches in diameter; central spines 3 (or sometimes 2), *all* bearing hooks, equal and similar, up to about ½ inch long, reddish-brown, hooked; radial spines tan, about 10 to 12, about ½ inch long, basally about 1/72 inch in diameter; petaloid perianth parts light reddish-purple; seeds obviously reticulate, the surface between the pits less prominent in the pattern than the pits.

DISTRIBUTION: Gravelly soils of plains and hills in grassland, desert grassland, and woodland at 5,000 to 8,000 feet elevation. Great Plains grassland and Juniper-Pinyon woodland. Arizona (according to report) near Springerville, Apache County (definitely collected just across the state line in New Mexico), in Pinal County between Superior and Sonora, and occasional southeastward to Cochise County. New Mexico from Zuñi in McKinley County to Sandoval, Guadalupe, Grant, and Doña Ana counties; Texas in the Franklin Mountains in El Paso County and at Samuels.

SYNONYMY: *Mammillaria Wrightii* Engelm. *Cactus Wrightii* Kuntze. *Neomammillaria Wrightii* Britton & Rose. *Chilita Wrightii* Orcutt. *Ebnerella Wrightii* F. Buxbaum. Type locality: Anton Chico, New Mexico. *Mammillaria Wilcoxii* Toumey. *Neomammillaria Wilcoxii* Britton & Rose. *Chilita Wilcoxii* Orcutt. *Mammillaria Wrightii* Engelm. var.

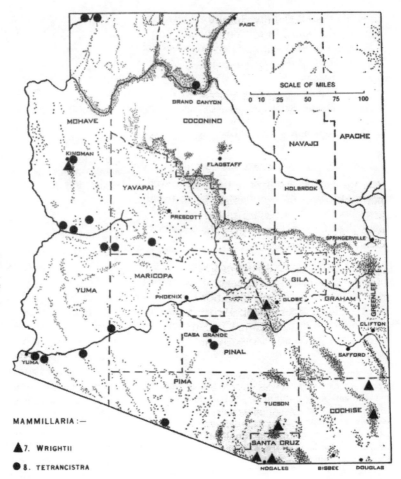

Fig. 4.14. Documented distribution of *Mammillaria Wrightii* and of *Mammillaria tetrancistra.*

MAMMILLARIA :—

▲ 7. WRIGHTII

● 8. TETRANCISTRA

Wilcoxii W. T. Marshall. *Ebnerella Wilcoxii* F. Buxbaum. Type locality: Fort Huachuca, Cochise County, Arizona. *Neomammillaria viridiflora* Britton & Rose. *Chilita viridiflora* Orcutt. *Mammillaria viridiflora* Bödeker. *Mammillaria Wilcoxii* Toumey var. *viridiflora* W. T. Marshall. *Mammillaria Wrightii* Engelm. *viridiflora* W. T. Marshall. Type locality: between Superior and Miami, Arizona.

8. *Mammillaria tetrancistra*
   Engelm.

Stem solitary, the larger plants ovoid-cylindroid to cylindroid, 3 to 6 inches long, 1½ to 2½ inches in diameter; spines very dense, obscuring the stem; principal (hooked) central spines 1 to 4 per areole, red to nearly black above, white near the bases; intermediate central spines none to several, straight, ranging from colored like the principal centrals to white, all the central spines spreading at various angles, the longer (hooked) ones ½ to 1 inch long, basally 1/64 to 1/60 inch in diameter; radial spines white or the larger ones tipped with red or red-to-black, intergrading with the centrals, 30 to 46 or reportedly 60 per areole, in about 2 or 3 (?) series, straight, up to ⅜ to ½ inch or 1 inch long, basally usually 1/120 inch in diameter; flower 1 to 1½ inches in diameter; petaloid perianth parts rose-pink to purple, oblanceolate, ⅝ to ¾ inch long, 1/12 to ⅛ inch broad, acute to short-acuminate-aristate, entire; fruit red, cylindroid to club-shaped, ¾ to 1¼ inches long, ¼ to 7/16 inch in diameter; seed black, dull, tuberculate, nearly a sphere, with the hilum invested by the nearly white corky tissue of the aril which is almost as large as the seed, the seed body 1/12 inch long (⅓ of it covered and obscured by the aril), the aril about 1/24 to 1/16 inch long in dried specimens, but, according to Engelmann (in Emory, Rept. U.S. & Mex. Bound. Surv. 2: Cactaceae

7. *pl. 7.* 1859) " . . . 1.2-1.5 line long . . ." (1 line being 1/12 inch or 2 mm.), this probably being based upon the size in living material before shriveling in drying. In living seeds the aril has 3 symmetrical inflated areas, as shown in Engelmann's illustration.

DISTRIBUTION: Sandy soils of hills, valleys, and plains in the desert at 450 to 2,400 feet elevation. Colorado Desert and the lower areas of the Mojavean and Arizona Deserts. Arizona in Mohave, Yuma, western Yavapai, western Maricopa, and western Pinal counties. California from Inyo County (Panamint Mountains) to San Diego and Imperial counties; southwestern Nevada; Utah near St. George.

SYNONYMY: Mammillaria tetrancistra Engelm. *Mammillaria phellosperma* Engelm., *nomen novum. Cactus phellospermus* Kuntze. *Cactus tetrancistrus* Coulter. *Phellosperma tetrancistra* Britton & Rose. *Neomammillaria tetrancistra* Fosberg. Neotype locality: Whitewater Canyon, Riverside County, California.

* * *

The proportionately immense aril developed as a 3-lobed structure about the hilum (attachment scar) of the seed is a feature of the species. Similar but less-developed structures occur in related species of *Mammillaria*. Presence of this single striking character led Britton and Rose to place the species in a genus *(Phellosperma)* by itself. This character, however, is not well correlated with others deviating from the combinations in *Mammillaria*.

9. *Mammillaria lasiacantha*
   Engelm.

Obscure, the stem projecting only slightly above the soil; stem top-shaped or turnip-shaped to nearly globose, ¾ to 1½ inches in diameter; spines obliterating the surface of the stem, white, in several undifferentiated series, about 40 or commonly 50 to 80 per areole, straight, about ⅛ inch long, acicular, white-pubescent to glabrous; flower rotate, ⅜ to ½ inch in diameter; petaloid perianth parts each white with a red midstripe, very narrowly elliptic, up to about ⅜ inch long, 1/12 inch broad; fruit red, fleshy but thin-walled, ultimately drying and shrivelling, cylindroid to clavate, ½ to ¾ inch long, ⅛ to 3/16 inch in diameter; seeds brown with a darker covering, reticulate-pitted, longer than broad, about 1/24 inch long.

DISTRIBUTION: Limestone hills and mesas in deserts and grasslands at 3,000 to 4,300 feet elevation. Chihuahuan Desert and perhaps the Desert Grassland. Cochise County, Arizona (according to report). South-central New Mexico; Texas near the Rio Grande as far eastward at the Big Bend. Inclusion of the species in this book is based upon statements by W. T. Marshall, Saguaroland Bulletin 6:32. 1952; 7:54. 1953; Arizona Cactuses, ed. 2. 102–103. 1953. Two plants are said to have been collected by Alan Blackburn on the Santa Cruz-Cochise county line, but no specimens have been found.

This species is confused readily with *Epithelantha micromeris.* In that species the flowers are produced on newly forming tubercles at the apex of the stem, rather than in a circle at some distance from it. In *Epithelantha* the new spines form a conspicuous tuft at the apex of the stem, but each disarticulates in the middle leaving a short stump. Consequently the areas away from the growing point are covered by short spines.

5. FEROCACTUS
Barrel Cactus

Stems unbranched or branched usually after injury to the terminal bud, cylindroid to ovoid or depressed-globose, usually ½ to 10 feet long, ⅙ to 1½ or 2 feet in diameter; ribs 13 to 30, the tubercles almost completely coalescent. Leaves not discernible on the mature tubercle. Areoles nearly circular to elliptic. Spines with cross-ribs or smooth, either red, pink, white, tan, brown, or yellow, the surface layer often ashy-gray or becoming gray in age; central spines 4 or rarely 1 or 8, straight, curved, or hooked, up to 6½ inches long, basally 1/48 to 1/6 inch diameter, needlelike or flattened, broadly to narrowly elliptic in cross section; radial spines either colored like the radial or usually lighter or white, 6 to 20 per areole, straight or curved, 3/8 to 3 inches long, basally 1/96 to 1/24 or 1/8 inch in diameter or width. Flowers and fruits on the new growth of the current season and therefore near the apex of the stem or branch, at the apex of the tubercle in a felted area adjacent to and merging with the spine-bearing part of the areole, this area persisting for many years after the fall of the fruit and forming a semicircular to circular scar. Flower

1½ to 3 inches in diameter; superior floral tube obconical to barely funnelform, green or tinted like the perianth. Fruit fleshy, with numerous or sometimes only 10 to 15 broad scales, these with scarious margins, fimbriate or denticulate, short-cylindroid, ovoid, or globular, ⅓ to 1¾ inches long, ⅓ to 1⅜ inches in diameter, with the floral tube persistent, opening by a short crosswise or lengthwise slit between the base and the middle. Seeds black, with a fine meshwork or with pits or papillae, narrowly compressed-obovoid to semicircular or obovoid with the base flaring around the micropyle, longer than broad (length is hilum to opposite side), 1/24 to 1/8 inch in greatest dimension; hilum obviously basal or "sub-basal."

DISTRIBUTION: Twenty or thirty species occurring from California to Texas, western Mexico, and the Mexican Plateau. Six in the United States. Three in Arizona.

LIST OF SPECIES AND VARIETIES

KEY TO THE SPECIES

1. Radial spines 12 to 20, the outer ones *either* (1) bristlelike, flexible, white, and irregular *or* (2) rarely, in a variety occurring from western Pima County to Gila County, Arizona, stiff, yellow, and merely curving; petaloid perianth parts at least partly yellow and with some red.

  2. The 4 central spines surrounded by a series of radial spines of similar texture, only the outer spines slender, flexible, and curving irregularly; plant flowering profusely in May or June following the winter rains and sometimes a little after the summer rains; principal central spine red to pink with the tip yellow or rarely straw-colored, *not* hooked or recurved at the apex (at least not so in most of the areoles of the plant; occasionally curved 90° but not recurved), 2 to 5½ inches long, 1/12 to 1/8 or rarely 1/6 inch broad; stigmas cylindroidal, about 1/48 inch in diameter; seed with the hilum clearly basal, the coat surface deeply reticulate-punctate.

      1. *Ferocactus acanthodes,* page 164

  2. The 4 central spines surrounded by no or only a few spines of similar texture, all or

nearly all the radial spines slender, flexible, and irregularly curving; plant flowering in late summer and sometimes a little in early summer; principal central spine red with a surface covering of ashy gray, hooked or recurved at the apex, 1½ to 2 inches long, 1/16 to 1/12 inch broad; stigmas flattened, 1/20 inch broad; seed with the hilum appearing sub-basal, the coat surface with a shallow meshwork.

      2. *Ferocactus Wislizenii,* page 166

1. Radial spines mostly 7 to 10, rigid, straight or slightly curving; spines *not* yellow; perianth parts purplish-red; plant flowering in early summer.   3. *Ferocactus Covillei,* page 169

1. *Ferocactus acanthodes*
(Lemaire) Britton & Rose

Stems columnar or sometimes barrel-shaped, and nearly always solitary, that is, unbranched, cylindroid and usually elongate (the younger ones sometimes ovoid-cylindroid), usually 3 to 10 feet long, about 1 foot in diameter; ribs mostly 18 to 27; tubercles nearly indistinguishable; spines dense, more or less obscuring the stem; central spines yellowish or red-and-yellow when young, usually turning red except apically, later often gray, usually 4 per areole in the form of a cross, the upper and lower ones broader, longer, and thicker, the lower usually apically curving a little, sometimes curving up to 90°, but not hooked, the longest up to 2 to 5½ inches long, basally 1/2 to 1/8 or 1/16 inch broad, the larger ones narrowly elliptic in cross section; outer 6 to 12 radials as long as the inner ones but about 1/48 inch in diameter, flexible, irregularly curving in and out; all but the slender outer radials (if any) strongly cross-ribbed; flower 1 1/2 to 2 1/2 inches in diameter; petaloid perianth parts yellow with some red along especially the basal portions of the veins, the largest oblanceolate, 3/4 to 1 3/4 inches long, 3/16 to 1/4 inch broad, acute to mucronate-acuminate, somewhat serrulate-dentate; fruit yellow, fleshy at maturity, with numerous rounded, scarious-margined scales, the thick wall usually 1¼ to 1½ inches long, about ⅝ to ¾ inch in diameter; seed minutely reticulate-pitted, 1/12 to 1/8 inch long.

Flowering occurs almost wholly in late spring or early summer following the winter rains, but sometimes sporadic flowering takes place at other times.

1A. Var. *acanthodes*
    (Table 15)

DISTRIBUTION: Gravel or rock of hills, canyon walls, alluvial fans, and wash margins in the desert at 200 to 1,500 or 2,000 feet elevation. Colorado Desert. Arizona near the lower Gila River. California in eastern San Bernardino, Riverside, San Diego, and Imperial counties. Mexico in adjacent Baja California.

SYNONYMY: Echinocactus acanthodes Lemaire. *Ferocactus acanthodes* Britton & Rose. Baja California. *Echinocactus viridescens* Nutt. var. *cylindraceus* Engelm. *Echinocactus cylindraceus* Engelm. *Ferocactus cylindraceus* Orcutt. Type locality: San Felipe, San Diego County, California. *Ferocactus Rostii* Britton & Rose. *Echinocactus Rostii* Berger. *Echinocactus acanthodes* Lemaire var. *Rostii* Munz. *Ferocactus acanthodes* (Lemaire) Britton & Rose var. *Rostii* Marshall & Bock. Type locality: Lower

California 40 miles south of the International Boundary.

1B. Var. *LeContei*
    (Engelm.) Lindsay
    (Table 15)

DISTRIBUTION: Gravel or rock of hills, canyon walls, alluvial fans, and wash margins or sometimes sandy flats in the desert at mostly 2,500 to 5,000 feet elevation or in central and southwestern Arizona 1,000 to 3,000 feet. Mojavean and Arizona deserts and the upper edge of the Colorado Desert. Arizona from Mohave County to Coconino (Grand Canyon and lower Little Colorado River up to about 4,500 feet elevation), Gila Yuma, and Pima counties (rare in eastern Pima County). California in the mountains of the eastern and southern Mojave Desert and the upper edge of the western Colorado

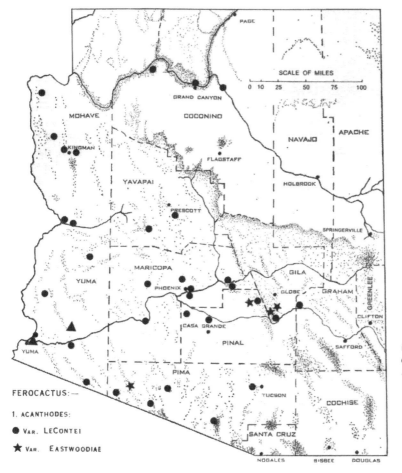

Fig. 5.1. The documented distribution of *Ferocactus acanthodes,* according to its varieties.

Fig. 5.2. A barrel cactus, *Ferocactus acanthodes* var. *LeContei,* common from Phoenix westward. Characteristically the plants are tall and columnar. These are seven feet high. (Converted from Kodachrome by Robert C. Frampton Studios, Claremont, California.)

Desert in Riverside and San Diego counties and rare in northeastern Imperial County; Nevada in southern Clark County; Utah probably near St. George, Washington County. Mexico in adjacent Sonora and probably northern Baja California.

SYNONYMY: Echinocactus LeContei Engelm. *Echinocactus Wislizenii* Engelm. var. *LeContei* Engelm. *Ferocactus LeContei* Britton & Rose. *Echinocactus LeContei* "Benson, et al, 1940" ex W. H. Earle, *pro. syn.,* incorrectly ascribed to the writer. *Ferocactus acanthodes* (Lemaire) Britton & Rose var. *LeContei* Lindsay. Type locality: Cactus Pass, headwaters of Bill Williams River, Arizona.

1C. Var. **Eastwoodiae**
L. Benson, **var. nov.** (p. 23)
(Table 15)

DISTRIBUTION: Mostly on inaccessible rocky ledges in the desert at 1,300 to 3,800 feet elevation. Arizona Desert. Arizona in western Pima County near the Organ Pipe Cactus National Monument and in the Dripping Springs and Mescal Mountains in Pinal and Gila counties.

SYNONYMY: Ferocactus acanthodes Lemaire var. *Eastwoodiae* L. Benson. Type locality: near Superior, Arizona.

2. *Ferocactus Wislizenii*
(Engelm.) Britton & Rose

Barrel-shaped to columnar plant, rarely with more than one stem (branching sometimes following injury), the stems massive, especially becoming so after damage to the terminal bud, cylindroid or in younger plants ovoid, 2 to 5 or 10 feet long, 1 or 1½ to 2 feet in diameter; ribs about 20 to 28, not markedly tuberculate; spines somewhat obscuring the stem; central spines red or the surface layer of ashy gray, 4 per areole, forming a cross, *not* flattened against the stem, strongly cross-ribbed, the lower ones or some of them hooked, up to 1½ to 2 inches long, basally 1/16 to 1/12 inch broad, narrowly elliptic in cross section; radial spines ashy gray, mostly 12 to 20 per areole, spreading, curving irregularly back and forth, not cross-ribbed, the longer ones up to 1¾ inches long, basally mostly 1/96 to 1/48 inch in diameter, needlelike; flower 1¾ to 2½ inches in diameter; petaloid perianth parts orange-yellow, narrowly lanceolate, 1 to 1½ inches long, about 3/16 inch broad, apically sharply acute and mucronate, marginally irregularly serrulate; fruit yellow, barrel-shaped, fleshy, covered by numerous almost circular, shallowly fimbriate scales, 1¼ to 1¾ inches long, 1 to 1⅜ inches in diameter; seed minutely and shallowly reticulate, 1/12 to 1/10 inch long.

DISTRIBUTION: Rock, gravel, or sandy soils of hills, canyons, washes, and alluvial fans in the desert or grassland at 1,000 or

Fig. 5.3. A yellow-spined barrel cactus, *Ferocactus acanthodes* var. *Eastwoodiae*. Characteristically this variety occurs on steep slopes or on cliffs. (Converted from Kodachrome by Robert C. Frampton Studios, Claremont, California.)

TABLE 15. CHARACTERS OF THE VARIETIES OF FEROCACTUS ACANTHODES

	A. Var. **acanthodes**	B. Var. **Le Contei**	C. Var. **Eastwoodiae**
Longest (lower) central spine	3 to 6 inches long, basally 1/12 or up to 1/8 inch broad, the apex curving up to about 90° but not recurved, at maturity red or becoming gray.	2 to 3 inches long, basally 1/12 or mostly 1/8 or sometimes 3/16 inch broad, the apex curving a little, at maturity red or becoming gray.	3 to 3¼ inches long, basally about 1/10 inch broad, the apex curving a little, conspicuously yellow or straw-yellow.
Radial spines	The inner 6 to 8 similar to the centrals but 1½ to 2½ inches long and 1/24 to 1/16 inch in diameter, the outer 6 to 12 1/48 inch in diameter, flexible, irregularly curving in and out, nearly white.	The inner 6 to 8 similar to the centrals but 1½ to 2 inches long and 1/24 to 1/16 inch in diameter, the outer 6 to 12 1/48 inch in diameter, flexible, irregularly curving in and out, nearly white.	All the 12 to 14 similar to the centrals but 1¾ to 2¼ inches long and 1/48 to 1/24 inch in diameter, all rigid and nearly straight, none curving in and out, conspicuously yellow or straw-yellow.
Petaloid perianth parts	3/16 to ¼ inch broad.	¼ to ⅝ inch broad.	About ⅜ inch broad.
Seed	About ⅛ inch broad.	About 1/12 inch broad.	1/12 inch broad (immature).
Geographical distribution	Colorado Desert. Arizona near the lower Gila River. California in eastern San Bernardino, Riverside, eastern San Diego, and Imperial counties. Mexico in northern Baja California.	Mojavean and Arizona deserts. Arizona from Mohave County to Gila and central Pima counties. Califorina in the higher parts of the eastern and southern Mojave Desert and the western Colorado Desert; Nevada in Clark County; Utah in Washington County.	Arizona Desert. Arizona in mountains of westernmost Pima County, southeastern Pinal County, and southwestern Gila County.

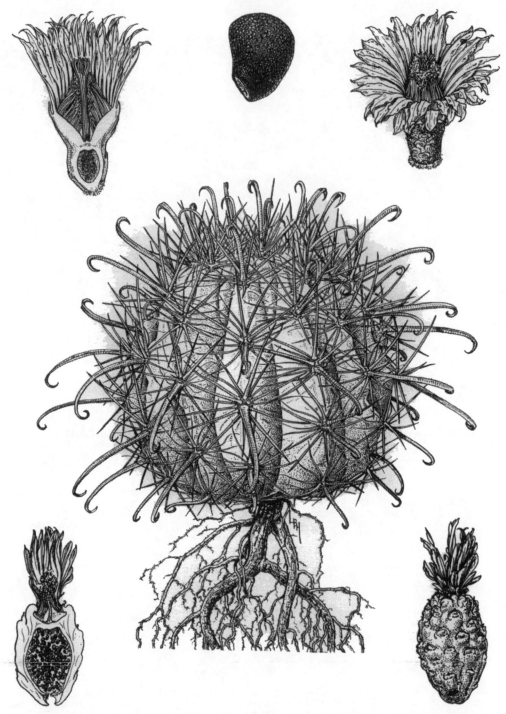

Fig. 5.4. A barrel cactus, *Ferocactus Wislizenii,* common from Phoenix and the Organ Pipe Cactus National Monument eastward. Characteristically the plants are stout and 2 to 5 feet high. The one here *(center)* is young and about 1 foot in both length and diameter; the length increases later but the diameter does not increase proportionately. *Above,* flower in longitudinal section and in external view and an enlarged seed. *Below,* fruits in longitudinal section and in side view.

Fig. 5.5. Barrel cacti. *Above, Ferocactus Wislizenii* in flower. Note the large hooked lower central spine, the lesser straight upper central spines, and the slender, flexuous, white radial spines. *Below, Ferocactus Covillei.* Note the single apically curving but (except in young plants) not usually hooked solitary central spine and the relatively large stout, inflexible radials. (Photographs by Robert A. Darrow.)

2,000 to 4,500 or (eastward) 5,600 feet elevation. Arizona Desert; Desert Grassland; Chihuahuan Desert. In areas with summer rainfall. Flowering profusely in the summer and sometimes sporadically in late spring or early summer. Southeastern Arizona from the vicinity of Phoenix and (mostly eastern) Pima County eastward. Southern New Mexico; extreme western Texas in El Paso County. Mexico southward to Sinaloa and northwestern Chihuahua.

The large yellow fruits, which ripen in early winter, are a favorite food for deer and rodents. The abundance of mammals in a district may be determined accurately by presence or absence of barrel cactus fruits in the late winter or spring.

SYNONYMY: Echinocactus Wislizenii Engelm. *Ferocactus Wislizenii* Britton & Rose. Type locality: Doñana, Doña Ana County, New Mexico. *Echinocactus Emoryi* Engelm. *Echinocereus Emoryi* Rümpler. *Ferocactus Emoryi* Backeberg. Type locality: Guthrie, Greenlee County, Arizona. *Echinocactus Wislizenii* Engelm. var. *decipiens* Engelm. Type locality: Bowie, Cochise County, Arizona. *Echinocactus Thurberi* Toumey, *nom. nud.* Type locality: east of Phoenix, Arizona. *Echinocactus Wislizenii* Engelm. var. *albispina* Toumey. *Ferocactus Wislizenii* (Engelm.) Britton & Rose var. *albispinus* Y. Ito. Type locality: foothills of the Santa Catalina Mountains, Pima County, Arizona. *Echinocactus arizonicus* Kunze. *Ferocactus arizonicus* Orcutt. Type locality: Pinal County, Arizona. *Echinocactus Wislizenii* Engelm. var. *phoeniceus* Kunze. *Ferocactus phoeniceus* Orcutt. *Ferocactus Wislizenii* (Engelm.) Britton & Rose var. *phoeniceus* Y. Ito. Type locality: Southwest of Phoenix, Arizona.

3. *Ferocactus Covillei*
    Britton & Rose

Stem barrel-shaped, ovoid, or sometimes columnar, unbranched except after injury to the terminal bud, 2 to 5 or up to 8 feet long, 1 to 1½ or 2 feet in diameter; ribs usually 20 to 30, not tuberculate; spines partly obscuring the stem; central spine red, but developing a surface layer of ashy gray, 1 per areole, standing at right angles to the stem, curving slightly through its whole length, commonly (in Arizona) hooked or strongly curved at the apex, strongly cross-ribbed, 3 to 4 inches long, basally up to ⅙ or ⅕ inch broad, flat on the upper surface, rounded below, semicircular in cross section; radial spines of the same texture and color as the central, 7 to 9 per areole,

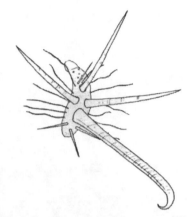

Fig. 5.6. A barrel cactus, *Ferocactus Wislizenii.* Spine cluster; 1 large hooked central spine; 3 lesser upper central spines; slender, flexuous, nearly white radial spines.

Fig. 5.7. *Ferocactus Wislizenii.* Fruit, showing the basal pore through which the seeds emerge after detachment of the fruit.

spreading, straight or slightly curving, up to 2 to 3 inches long, basally 1/12 to 1/8 inch broad, somewhat flattened, elliptic in cross section; flower about 3 inches in diameter; petaloid perianth parts purplish-red, narrowly lanceolate, about 1 1/2 inches long, about 3/16 to 1/4 inch broad, apically short-acuminate, fimbriate-serrulate; fruit yellow, with relatively few (25 to 30 or 40) semicircular, scarious-margined, fimbriate or crenulate scales, 1 to 1¾ inches long, 1 to 1¼ inches in diameter; seed surface with a meshwork, the seed 1/10 inch long.

DISTRIBUTION: Gravelly, rocky, or sandy soils of hills, washes, alluvial fans, or grassy flats in the desert at 1,500 to 2,500 or 3,000 feet elevation. Arizona Desert and upper Colorado Desert (in the United States). Arizona from (according to A. A. Nichol)

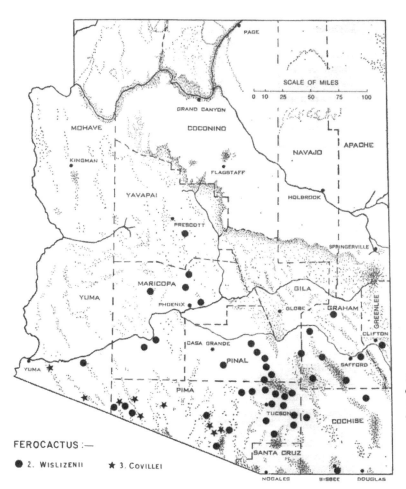

Fig. 5.8. The documented distribution of *Ferocactus Wislizenii* and of *Ferocactus Covillei.*

FEROCACTUS :—

● 2. WISLIZENII   ★ 3. COVILLEI

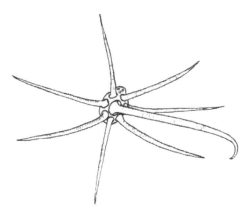

Fig. 5.9. A barrel cactus, *Ferocactus Covillei.* Spine cluster. Note the solitary central spine, in this case with a hook, though usually not so in mature plants, and the half dozen stout, rigid radial spines.

the Sierra Estrella, Maricopa County, to the Organ Pipe Cactus National Monument and the Papago Indian Reservation, Pima County. Mexico in adjacent Sonora.

This plant is perhaps the most distinctive of the Arizona barrel cacti or bisnagas. Large plants are tall and cylindrical. When the plant is young it is enclosed in a "cage" of bright red interlacing spines, which are recurved in such a manner that the cactus may be picked up without danger of injury. The spines form an almost perfect sphere around the young plant, and an unrooted specimen will roll and bounce down a hillside like a tennis ball. The armor of spines protects the young cactus from rodents and other plant-eating animals. The spines of young plants are remarkably variable.

SYNONYMY: Ferocactus Covillei Britton & Rose. *Echinocactus Covillei* Berger. Type locality: Altar, Sonora, Mexico. *Echinocactus Hert[r]ichii* Weinberg. *Thelocactus Hertrichii* Borg. Type locality: ". . . Tortilla and Gila mountains, Arizona." (Possibly this species.)

6. ECHINOCACTUS
Barrel Cactus

Stems either branching or unbranched, ⅙ to 2 feet long, ⅙ to 1 foot in diameter; ribs 8 to 27; tubercles almost completely coalescent, in some species with slits between the upper portions. Leaves not discernible in the adult plant. Spines, when present, annulate; central spines red, sometimes covered partly with ashy gray, 1 to 4 per areole, straight or curving, 1 to 3 inches long, basally 1/16 to ⅜ inch broad, needlelike or flattened; radial spines like the central but smaller, 5 to 11 per areole, ¾ to 2 inches long, basally 1/24 to ⅛ inch broad. Flowers and fruits on the new growth of the current season, therefore near the apex of the stem or branch, growing at the apex of the tubercle in a felted area adjacent to and merging with the spine-bearing part of the areole, this area persisting for many years and forming a circular or semicircular scar. Flower 1½ to 2¾ inches in diameter; floral tube above the ovary obconical to slightly funnelform, green or tinted like the perianth. Fruit dry or at first fleshy, with scales and with the hairs from beneath these sometimes obscuring the fruit, ⅝ to 2 inches long, ⅜ to 1½ inches in diameter. Seeds black or brown, reticulate, broader than long (length being hilum to opposite side), 1/12 to 1/6 inch in greatest dimension; hilum appearing "lateral" or oblique.

DISTRIBUTION: Twelve or more species occurring from California to Texas and southward into Mexico as far as Querétaro. Four in the United States. Two in Arizona.

LIST OF SPECIES AND VARIETIES

KEY TO THE SPECIES

1. Radial spines on all sides of the areole; central spines 3 or usually 4, spreading or curving in various directions, at maturity red or yellow or pale gray tinged with red; flowers yellow; fruit ovoid; stem ribs 13 to 21, nearly vertical, usually not markedly spiral; stems several to many, forming dense clumps, elongate-ovoid.
1. *Echinocactus polycephalus,* page 172

1. Radial spines lacking on the extreme lower side of the areole; principal central spine turned strongly downward, at maturity dark or dirty gray; flower pink; fruit narrowly cylindroid; stem ribs 7 to 13, commonly 8, markedly spiral, stem solitary, spheroidal to elongate-ovoid; stem solitary.
2. *Echinocactus horizonthalonius,* page 173

1. *Echinocactus polycephalus*
   Englem. & Bigelow

Stems forming clumps of up to 30 branches, the clumps up to 2 feet high and 4 feet in diameter; stems spheroid to cylindroid, 6 to 12 or 24 inches long, up to 8 to 10 or 12 inches in diameter; ribs 13 to 21; spines obscuring the stem; central spines red with an ashy surface layer or a dense felty canescence which may peel off in sheets as the spine ages, 4 per areole, spreading irregularly, the lower (principal) one slightly curving downward, the others nearly straight, strongly cross-ribbed, up to 2½ to 3 inches long, basally 1/10 to 1/8 inch broad, flattened; radial spines similar to the central but smaller, 6 to 8 per areole, irregularly spreading or curving slightly in low arcs, 1¼ to 1¾ inches long, basally 1/24 to 1/16 inch broad, somewhat flattened; flower 2 inches in diameter; petaloid perianth parts yellow or the midribs tinged with pink, oblanceolate, about ¾ inch long, about ⅛ inch broad, acute to mucronate or aristulate, slightly and minutely toothed; fruit dry, densely encased in matted woolly white hairs ½ to ¾ inch long, the ovary ¾ to 1 inch long, ⅜ to ½ inch in diameter; seed black, strongly reticulate, 1/10 inch long, 1/6 inch broad.

1A. Var. *polycephalus*

Stems numerous, forming clumps of 10 to 20 or even 30, the longest in the middle; spines densely ashy-canescent, the felt peeling away in sheets after maturing of the spine, the spine red or pink with an ashy covering even after disappearance of the felt, straight or somewhat curving.

DISTRIBUTION: Rocky or gravelly soils of slopes or sometimes the clay soils of the valleys in the deserts at 100 to 2,500 feet elevation. Mojavean Desert. Arizona from western Mohave County to Yuma County. California in the Mojave Desert and northeastern Imperial County; southern Nevada. Mexico in northwestern Sonora.

Fig. 6.1. A barrel cactus, *Echinocactus polycephalus*. A cluster of stems of the same individual, some with the dry fruits enveloped in the long white hairs from the areoles on their surfaces. (Photograph by Robert A. Darrow; converted from Kodachrome by Robert C. Frampton Studios, Claremont, California.)

SYNONYMY: Echinocactus polycephalus Engelm. & Bigelow. Type locality: Mojave River, California. *Echinocactus polycephalus* Engelm. & Bigelow var. *flavispina[us]* Haage fil. Without reference to a specimen or a locality.

1B. Var. *xeranthemoides*
   Coulter

Stems few, mostly 1 to 12 (according to A. A. Nichol) the clumps commonly asymmetrical, that is, with a marginal stem the tallest (the individuals observed with too few stems to confirm or reject this statement); spines not hairy or the hair falling away, yellow or tinged with red or pink or light red, straight or somewhat curving, though probably less so than in var. *polycephalus*.

DISTRIBUTION: Rocky, south-facing ledges of canyons and hills in the desert or woodland at 3,800 to 5,000 feet or more elevation. Navajoan Desert and the edge of the Juniper-Pinyon Woodland. Arizona in easternmost Mohave County and northern Coconino County near the Colorado River and its tributaries, such as Kanab Creek and the Little Colorado River. Nevada in Deadman Canyon, Clark County; Utah in the area about Kanab.

According to Britton and Rose (Cactaceae 3:167. 1922), the seeds of var. *polycephalus* are "papillate," those of var. *xeranthemoides* "smooth and shining." Those of var. *polycephalus* are reticulate; those of var. *xeranthemoides* remain to be studied.

SYNONYMY: Echinocactus polycephalus Engelm & Bigelow var. *xeranthemoides* Coulter. *Echinocactus xeranthemoides* Engelmann ex Rydberg. Type locality: Kanab Wash near the Colorado River, northern Arizona.

2. *Echinocactus horizonthalonius*
   Lemaire
   Turk's head

Stem solitary, depressed-globose to ovoid or columnar, blue-green, usually 4 to 6 but rarely 12 inches long, 4 to 6 inches in diameter; ribs 7 to 13, commonly 8; spines dense, more or less obscuring the stem, glabrous;

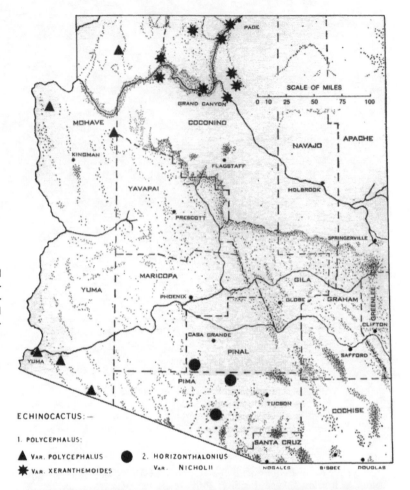

Fig. 6.2. The documented distribution of *Echinocactus polycephalus,* according to its varieties, and of *Echinocactus horizonthalonius* var. *Nicholii.*

SCALE OF MILES

0 10 25 50 75 100

ECHINOCACTUS:—

1. POLYCEPHALUS:

▲ VAR. POLYCEPHALUS

✸ VAR. XERANTHEMOIDES

● 2. HORIZONTHALONIUS VAR. NICHOLII

Fig. 6.3. *Echinocactus horizonthalonius* var. *Nicholii.* A dry, woolly fruit is at the stem-apex. (Converted from Kodachrome by Robert C. Frampton Studios, Claremont, California.)

central spine pale gray or sometimes black, this covering under-layers of red or red-and-yellow, 3 per areole, 1 curving gradually downward in a low arc, 1 to 1¼ inches long, basally 1/12 to 1/8 inch broad, somewhat flattened, 2 more slender upper ones curving upward or straight; radial spines gray, this color covering lower layers of red or yellow, 5 or 6 to 8 per areole, slightly curving outward in low arcs, ¾ to 1 inch long, basally about 1/24 to 1/16 inch in diameter, a little flattened; flower 2 to 2¾ inches in diameter; petaloid perianth parts pink, obovate-spathulate, 1 to 1¼ inches long, up to ½ inch broad, apically rounded, mucronate, and irregularly toothed, marginally entire; fruit dry, densely covered with white, soft, woolly hairs, about 1 inch long, ½ inch in diameter; seed black, with a fine surface meshwork, 1/12 to 1/7 inch long, 1/12 to 1/8 inch broad.

2A. Var. **Nicholii**
L. Benson, **var. nov.** (p. 23)

Stem becoming columnar in age, up to 1½ feet high, 5 or 6 or 8 inches in diameter; some spines tending to be nearly black or partly gray with under-layers of red; central spines 3, 1 black and curving downward, 2 red or basally pale gray and curving upward; radial spines 5; seed longer than broad, ⅛ to 1/7 inch long, ½ inch broad; hilum "sub-basal."

DISTRIBUTION: In the desert at about 3,000 to 3,500 feet elevation. Arizona Desert. Arizona in southwestern Pinal County and northcentral Pima County.

Knowledge of this variety is attributed primarily to Mr. A. A. Nichol, formerly of the University of Arizona and then of Arizona State Game and Fish Department. In the 1930's, Mr. Nichol considered the Arizona plants different from those occurring in Texas.

7. SCLEROCACTUS

Stems solitary or rarely branching (perhaps due to injury of the terminal bud), cylindroid, ovoid, globose, or depressed-globose, 2 to 8 or 16 inches long, 1½ to 4 or 6 inches in diameter; tubercles coalescent into the ribs through one-half to four-fifths their height; leaves not discernible; spines smooth, not with crosswise ridges; central spines gray, white, yellow, red, or brown, 1 to 6 or 11 or sometimes 0 per areole, *usually* of two or three distinctive types, *usually* 1 or more hooked, ½ to 3½ inches long, needlelike or flattened or commonly some of each type; radial spines white or gray or sometimes some of them pink or brown, 6 to 11 or 15 per areole, straight, shorter than the central spines, needlelike or flattened. Flowers and fruits of the new growth of the current season near the apex of the stem or branch, developed on the side of the tubercle toward the stem-apex in a felted area adjacent to and merging with the new spine-bearing areole, the scar persisting for many years after the fall of the fruit and forming a circular to irregular or sometimes an oblong scar. Flower ¾ to 2¼ inches in diameter; superior floral tube short, funnelform; petaloid perianth parts white, yellow, or from pink to reddish-purple. Fruit green, thin-walled, turning reddish and becoming dry, naked or with a few broad and thin scales, 3/8 to 1 inch long, 3/8 to 7/16 inch diameter, with the apical cup obscured by the persistent superior floral tube and flower parts, opening *either* along a circular nearly horizontal regular to somewhat irregular line at just above the base to below or even a little above the middle *or* (in one species) along 2 or 3 short vertical lines. Seed black, papillate-reticulate, broader than long, 1/16 to 1/10 inch long (from hilum to opposite side).

DISTRIBUTION: Six species from the Mojave Desert in California to the Great Salt Lake area in Utah and to the Colorado Plateau as far east as western Colorado and northwestern New Mexico.

LIST OF SPECIES AND VARIETIES

KEY TO THE SPECIES

1. Fruit opening horizontally near the base; spines of juvenile stems not densely white-pubescent, sometimes finely canescent (with very short hairs); only one (upper) central spine white, broad, and basally flattened (rarely the upper central only slightly flattened and pale instead of white); style finely pubescent with hooked hairs; scales on the inferior floral cup covering the ovary markedly membranous-bordered and tending to be fringed.
     1. *Sclerocactus Whipplei,* page 176

1. Fruit opening along two or three short vertical slits; the spines of the juvenile stems (those often still discernible on the bases of older stems) densely white-pubescent (bearing well-developed white hairs); *two or more* upper or lateral central spines, in at least some of the areoles of mature plants, white, broad, and basally flattened; style glabrous; scales on the inferior floral tube covering and adherent to the ovary only slightly membranous-bordered and *not* fringed.
     2. *Sclerocactus pubispinus,* page 178

1. *Sclerocactus Whipplei*
   (Engelm. & Bigelow) Britt. & Rose

Stems solitary or sometimes 2 or 3, elongate and cylindroid or the younger ones ovoid, up to 8 or 16 inches long, 4 or 6 inches in diameter, one plant (according to Donald G. Davis) weighing as much as 17 pounds); central spines mostly 4, the upper one white, nearly straight, flattened or narrowly elliptic or rhombic in cross section, 1 to 2 inches long, 1/24 to 1/8 inch broad, erect; radial spines mostly 7 to 11, all but two of the lower white, these like the lateral centrals with which they form a cross, the lower one reddish to purplish, pink, or tan, cylindroidal or somewhat angled, hooked, 1 to 2 inches long, about 1/24 inch in diameter, turned or curving more or less downward, the two lateral ones similar but a little shorter and not hooked, mostly about 1 inch long, basally 1/24 inch in diameter, elliptic or rhombic in cross section; flower 1 to 2¼ inches in diameter, 1 to 2⅝ inches long; petaloid perianth parts rose to purple, yellow, pink, or rarely white, oblanceolate, 1 to 2 inches long, ¼ to ½ inch broad, mucronate or short-setose, apically irregularly slightly toothed; fruit green, tan, or tinged with pink, dry, bearing a few scarious-margined, membranous-fringed scales, 1/2 to 3/4 or 1 inch long, 1/4 to 7/16 inch in diameter, opening irregularly along a

TABLE 16. CHARACTERS OF THE VARIETIES OF SCLEROCACTUS WHIPPLEI

	A. Var. roseus	B. Var. **intermedius**	C. Var. **Whipplei**
Stems	Elongate, cylindroidal, mostly 3 to 5 or 6 inches long, 2 to 2½ inches in diameter.	Elongate, cylindroidal, mostly 3 to 8 inches long, 2 to 2½ or 4 inches in diameter.	Stems depressed-globose to short-ovoid, up to 3 inches long or rarely longer, up to 3½ inches in diameter.
Upper central spine	Pale pink, somewhat flattened or angled, elliptic or rhombic in cross section, about 1/24 inch broad.	White, flat, narrow but ribbonlike, basally 1/24 to 1/16 inch broad, subulate.	White, flat, ribbonlike, conspicuous, basally 1/16 to ⅛ inch broad, subulate.
Lower (hooked) central spine	Reddish, slender, usually 1½ to 1¾ or 2 inches long.	Reddish, usually 1½ to 1¾ or 2 inches long.	Purplish-pink, up to 1 or 1½ inches long.
Flower	¾ to 1 inch in length and diameter.	1 to 2¼ inches in diameter, 1 to 2⅝ inches long.	About 1 inch in length and breadth.
Petaloid perianth parts	Purple, pink, or white.	Purple, pink, or white.	Yellow (so far as known).
Geographical distribution	Mostly Navajoan Desert, 3,500 to 5,000 or 6,700 feet elevation. Mostly near the large rivers. Middle Arizona near the Colorado River. Southeastern Utah and southwestern Colorado.	Mostly Juniper-Pinyon Woodland, 3,500 or 5,000 to 7,000 feet elevation. Arizona along the northernmost edge. Eastern Utah western edge of Colorado.	Navajoan Desert, 5,000 to 6,000 feet elevation. Arizona on the Little Colorado River drainage in Navajo and Apache counties.

Fig. 7.1. *Sclerocactus Whipplei* var. *intermedius* in flower. This is the common variety. Note the broad, flat, white upper central spine, visible in each upper areole. (Photograph by Robert H. Peebles.)

horizontal cleft at or above or below the middle; seeds 1/12 inch long (hilum to opposite point), 1/8 inch broad.

The species is variable. It may include more varieties than are listed here. Plants from near the "Four Corners" country of Utah, Colorado, Arizona, and New Mexico are being studied, but these are too little known to determine whether they represent taxa.

1A. Var. *roseus*
(Clover) L. Benson

DISTRIBUTION: Disintegrated red sandstone and other sandy or gravelly soils in the desert at 3,500 to 5,000 or 6,700 feet elevation. Lower Navajoan Desert and therefore almost restricted to the areas near the major watercourses, such as the Fremont, Colorado, Green, and San Juan rivers in southeastern Utah and westernmost Colorado and northern

Arizona; occurring mostly from the vicinity of Capitol Reef, Moab, and Mexican Hat downstream to Havasupai Canyon and southeastward to Kayenta.

SYNONYMY: Sclerocactus parviflorus Clover & Jotter. *Echinocactus parviflorus* L. Benson. Type locality: Forbidding Canyon, adjacent to the Canyon of the Colorado River, Utah. *Sclerocactus havasupaiensis* Clover. *Echinocactus parviflorus* (Clover & Jotter) L. Benson var. *havasupaiensis* (Clover) L. Benson. Type locality: Havasupai Canyon, Coconino County, Arizona. *Sclerocactus havasupaiensis* Clover var. *roseus* Clover. *Echinocactus parviflorus* (Clover & Jotter) L. Benson var. *roseus* (Clover) L. Benson. *Sclerocactus Whipplei* (Engelm. & Bigelow) Britton & Rose var. *roseus* (Clover) L. Benson. Type locality: Havasupai Canyon, Coconino County, Arizona.

1B. Var. *intermedius*
(Peebles) L. Benson

DISTRIBUTION: Gravelly or sandy soils in the desert or mostly the juniper-pinyon belt at 3,500 or 5,000 to 7,000 feet elevation; often on red sandstone soils. Juniper-Pinyon Woodland and the upper parts of the Navajoan Desert. Colorado River drainage from eastern Utah to the western edge of Colorado, northernmost Arizona, and San Ysidro, New Mexico. This is the *common variety,* long supposed to be var. *spinosior (Sclerocactus pubispinus).*

A few plants from the Colorado Plateau have yellow spines, and occasionally one with curving (but not hooked) central spines has been reported. No authentic specimens with records of the localities of collection have been found.

SYNONYMY: Sclerocactus intermedius Peebles. *Sclerocactus Whipplei* (Engelm. & Bigelow) Britton & Rose var. *intermedius* (Peebles) L. Benson. Type locality: west of Pipe Spring, Mohave County, Arizona.

1C. Var. *Whipplei*

DISTRIBUTION: Gravelly hills and canyons in the desert at 5,000 to 6,000 feet elevation. Lower Navajoan Desert. Arizona on the Little Colorado River drainage in Navajo and Apache counties.

A remarkably large living plant with the spines not hooked was seen in 1965.

SYNONYMY: Echinocactus Whipplei Engelm. & Bigelow. *Sclerocactus Whipplei* Britton & Rose. Type locality: Petrified Forest east of Holbrook, on Lithodendron Wash, Apache County, Arizona. *Sclerocactus Whipplei* (Engelm. & Bigelow) Britton & Rose

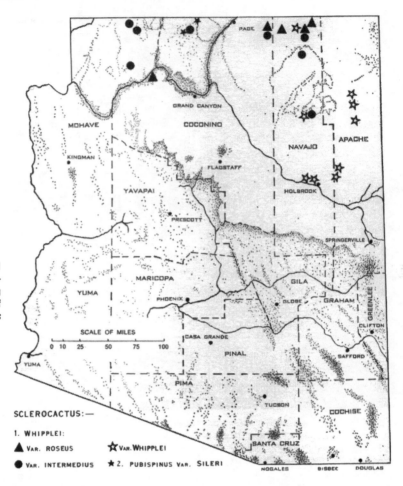

Fig. 7.2. The documented distribution of *Sclerocactus Whipplei,* according to its varieties, and of *Sclerocactus pubispinus* var. *Sileri.*

SCLEROCACTUS:—

1. WHIPPLEI:

▲ Var. ROSEUS   ☆ Var. WHIPPLEI

● Var. INTERMEDIUS   ★ 2. PUBISPINUS Var. SILERI

var. *pygmaeus* Peebles. Type locality: north of Ganado, Apache County, Arizona. This was a juvenile plant.

2. *Sclerocactus pubispinus*
(Englem.) L. Benson

Stem depressed-globose to ovoid, solitary, 1½ to 3 inches long, 2 to 2½ inches in diameter; spines rather dense, obscuring the stem, those of juvenile plants densely white-pubescent, those of the mature stems glabrous, the central spines white on juvenile stems and therefore on the older (lower) parts of mature stems (central spines described here for var. *pubispinus;* compare description for var. *Sileri* below); lower central spines of mature areas brown, the upper ones white, about 6 per areole, the principal lower one and sometimes one or two others hooked, otherwise straight, the lower central up to 1¼ inches long, basally about 1/24 to 1/20 inch broad, more or less needlelike but elliptic in cross section; upper median central spine white, basally flattened, up to 2¼ inches long, up to 1/16 inch broad, the 1 or 2 adjacent (probably central) spines similar but smaller; radial spines white, as many as 8 per areole, spreading, flexible, nearly straight, up to 1 inch long, slender, basally 1/96 to 1/48 inch broad, needlelike or flattened; flower about 1 inch in diameter; petaloid perianth parts reddish-purple, oblanceolate, about 1 inch long, 1/4 to 5/16 inch broad, mucronate, nearly entire; fruit green, becoming reddish and dry, with a few scales and these somewhat scarious on the margins and the hairs in the areoles beneath them short, barrel-shaped, about 3/8 inch long, of the same diameter; seed 1/12 to 1/10 inch long (hilum to point opposite), 1/8 to 1/6 inch broad.

2A. Var. **Sileri**
   L. Benson, **var. nov.** (p. 23)

Spines nearly all white, the bases of some or all of a few brown; lower hooked central spine and the two lateral centrals flattened, but narrower and thicker than the middle uppermost central; lower (principal) hooked central spine markedly longer than the others.

DISTRIBUTION: Upper edge of the Navajoan Desert. Northern Arizona in Houserock Valley, Coconino County. "Southern Utah," doubtless near Kanab and possibly along the area north of the Kaibab Plateau in Arizona. The distinction of the state boundary may not have been made with clarity at the time of collection, and the records in herbaria may represent guesses at localities of collection based upon the place of residence of the collector.

Because of the numerous white, flattened spines of the only collection of Var. *Sileri* known before 1963, various authors supposed this plant to be perhaps *Pediocactus papyracanthus* (of northern Arizona and northern New Mexico). See Britton & Rose, Cactaceae 3:92. 1922, where the *Siler* collection is described.

* * *

Sclerocactus mesae-verdae (Boissevain ex Hill & Salisbury) L. Benson. Mesa Verde cactus. (*Coloradoa mesae verdae* Boissevain, Colorado Cacti 55. 1940; *Coloradoa mesae-verdae* Boissevain ex Hill and Salisbury, Index Kewensis Suppl. 10:57. 1947. *Echinocactus mesae-verdae* L. Benson.) Small cacti usually with solitary stems; stem 2 to 2½ inches high; 3 to 3½ inches in diameter, ribbed, the ribs formed by confluence of the tubercles; ribs 13 to 17; areoles with brown or (later) gray wool; central spines usually none, rarely one, this straight or sometimes hooked; radial spines 8 to 10, about ⅜ inch long, eventually curved, flattened in age, spreading parallel to the stem, gray in age; flowers about 1½ inches in diameter; sepals purplish brown; petals cream to yellow with brownish midribs; fruit not scaly or hairy, perianth persistent, irregularly dehiscent crosswise.

DISTRIBUTION: South slopes of very dry hills. Mesa Verde cliffs in the southwestern corner of Colorado; Shiprock region, New Mexico. To be expected in Utah and Arizona in the vicinity of the Four Corners, since areas of collection are only 10 to 15 miles from the borders of both these states.

8. PEDIOCACTUS
Britton & Rose

Stems solitary or sometimes several, ½ to 3 or 6 inches long, ⅜ to 3 or 6 inches in diameter, cylindroid to globose or depressed-globose; ribs none; leaves none or microscopic. Spines more or less obscuring the joint, exceedingly variable in color, number per areole, direction, texture, and flattening, 1/24 or 1/4 to 1/2 or 1 1/4 inches long, basally 1/240 or 1/72 to 1/24 inch in diameter or breadth, needlelike or in one species flattened. Flower ⅖ to 1 inch in diameter; petaloid perianth parts white or with pink or yellow on at least the midribs, oblanceolate, obovate-oblanceolate, or narrowly oblong-cuneate, ⅙ or 1/2 to 3/4 or 1 inch long, 1/16 or 3/16 to 1/4 inch broad, acute to rounded, entire or mucronate or shallowly indented, marginally entire or minutely denticulate; fruit green, often changing toward tan or yellow, dry, naked or with a few scale-leaves, otherwise smooth except for the veins, obovoid to subglobose or cylindroid but tending to enlarge upward, 1/6 to 5/8 inch long, 3/16 to 5/16 inch in greatest diameter, with the apical cup shallow, both circumscissile on the broad apical margin and dehiscent along a dorsal slit produced by another abscission layer. Seeds black or gray, papillate-tessellate, broader than long (length being hilum to opposite point), 1/24 to 1/8 inch long, 1/16 to 3/16 inch broad; hilum appearing "lateral."

DISTRIBUTION: Seven species occurring in the mountainous portions of the Columbia River Basin and the Great Basin, in the Rocky Mountains, and on the Colorado Plateau, where six species and their varieties are restricted endemics.

Aside from remarkable variations in the spines, the seven species occurring in the western United States form a coherent unit. The gaps between spine types are becoming less marked as new species and varieties come to light. Mexican species related to this genus are in need of study.

Pediocactus is not well known because most species are small, of limited geographical occurrence, and restricted to areas until recently little visited. Some are of no greater diameter at maturity than an American twenty-five or fifty-cent piece. They may rise no higher above ground than the flat coin. Finding them may require days of searching even though the approximate location is known. In the dry season some species retract under the desert soil or gravel, limiting the period in which they may be found. Most are hidden in gravel or rocks of a special kind, and *Pediocactus papyracanthus* may grow in the fairy rings of blue grama grass, the dried leaves of which almost duplicate the appearance of the cactus spines.

Five new taxa, *Pediocactus Paradinei, Pediocactus Knowltonii, Pediocactus Peeblesianus* vars. *Fickeiseniae* and *Maianus*, and *Pediocactus Bradyi,* have been discovered within the last ten years. Recent discovery of such a high percentage of conservatively interpreted species and varieties is rare in the genera of plants occurring in the United States.

LIST OF SPECIES AND VARIETIES

KEY TO THE SPECIES

1. Spines *not* strongly flattened, needlelike, circular to elliptic in cross section; stems globular, depressed-globular, obovoid, or short-cylindroid, their length little greater than their diameter or rarely twice as great.
  2. Surface of the spine smooth, often more or less polished, rarely finely canescent.
    3. Sepaloid perianth parts and the few (if any) scales on the superior floral tube either minutely toothed or short-fimbriate or entire and often undulate; seed black, 1/16 to 1/8 inch long; petaloid perianth parts pink and white, white, magenta, or yellow; areole not more than ⅛ inch in diameter; spines slender, not more than 1/32 inch in diameter.
      4. Central spines none or, if (commonly) present, rigid, gently curving or straight, in mature plants at least distally reddish-brown or reddish, 5/16 to 1/2 or 1 1/16 inches long, 1/72 or 1/48 to 1/32 inch in diameter; petaloid perianth parts marginally either pink or magenta or white with pink middles or wholly yellow.
        5. Central spines present (except in juvenile plants or the lower areoles persisting on adult stems), straight, 5 to 8 or 11 (or in young plants as few as 3) per areole; ovary with a few scales; radial spines almost straight, spreading irregularly, ¼ to ⅜ or ¾ inch long; stems 1 to 5 or 6 inches long, 1 to 4 or 5 inches in diameter; scales of the floral tube

toothed or often short-fimbriate; seed about 1/12 inch long; fruit not stalked; seed tessellate-tuberculate.
        1. *Pediocactus Simpsonii,* page 180
        5. Central spines none (or rare); ovary practically lacking scales; radial spines slightly recurved, like the teeth of a comb along the elliptic or elongate areole, ½ to ¼ inch long; stems at maturity only 1 to 2 or 2½ inches in diameter, often barely protruding above ground; scales of the floral tube minutely toothed; fruit basally constricted into a short stalk; seed papillate and with larger mounds on the surfaces.
        2. *Pediocactus Bradyi,* page 181
      4. Central spines flexible and hairlike, bending or curving irregularly or straight; uniformly colored, white or ashy gray, turning in age to straw- or cream-color, 1 to 1 5/16 inches long, about 1/96 to 1/72 inch in diameter; petaloid perianth parts white or with pink midribs.
        3. *Pediocactus Paradinei,* page 181
    3. Sepaloid perianth parts and the scales of the floral tube long-fimbriate; seed gray, 1/16 to 1/5 inch long; petaloid perianth parts yellow or yellow with maroon veins; areole about ¼ inch in diameter; spines rather stout, 1/32 to 1/24 inch in diameter.
        4. *Pediocactus Sileri,* page 183
  2. Surface of the spine and the tissue beneath spongy-fibrous; sepaloid perianth parts and the scales of the ovary, when present, scarious-margined, never fimbriate.
        5. *Pediocactus Peeblesianus,* page 184
1. Spines strongly flattened, several times broader than thick, puberulent; stems elongate, their length at least twice their diameter.
        6. *Pediocactus papyracanthus,* page 186

1. *Pediocactus Simpsonii*
   (Engelm.) Britton & Rose

Stems solitary or sometimes a few in a cluster, globular to depressed-ovoid, 1 to 5 or 8 or even 12 inches long, usually 1 to 5 or 8 inches in diameter; central spines reddish-brown or sometimes red or basally pale yellow to cream, 5 to 8 or 11 (or in young plants none or as few as 3) per areole, spreading, straight, ⅜ to ½ or ¾ inch long, basally 1/72 inch in diameter, needlelike; radial spines white to cream, 15 to 30 per areole, spreading at right angles to the tubercles, nearly straight, ¼ to ⅜ inch long; flower ¾ to 1 inch in diameter; petaloid perianth parts pink to nearly white or sometimes magenta or yellow, narrowly oblong-cuneate, 1/2 to 3/4 or 1 inch long, 3/16 to 1/4 or 3/8 inch broad, mucronate or cuspidulate, apically irregularly shallowly indented; fruit greenish, tinged with red, dry, smooth, with a few scales on the upper portion, short-cylindroidal but bulging irregularly and widened upward, ¼ to ⅜ inch long, 3/16 to 5/16 inch in diameter, with the apical cup depressed and with a thin border; seeds gray or nearly black, tessellate-tuberculate, about 1/16 inch long, up to more than 3/32 inch broad.

DISTRIBUTION: Fine powdery soils of valleys and hills in dry areas at 6,000 to 9,500 feet elevation. Juniper-Pinyon Woodland and Sagebrush Desert. Northern Arizona (Grand Canyon). Eastern Oregon; southern Idaho to southern Wyoming and westcentral Nevada, northern New Mexico, westernmost South Dakota, and reportedly Montana and western Kansas.

SYNONYMY: Echinocactus Simpsonii Engelm. *Mammillaria Simpsonii* M. E. Jones. *Pediocactus Simpsonii* Britt. & Rose. Type locality: Kobe Valley, Nevada. *Mammillaria Purpusii* K. Schum. Type collection: Grand Mesa, Delta County, Colorado. *Pediocactus Simpsonii* (Engelm.) Britton & Rose var. *caespiticus* Backeberg. *nomen nudum.* Type locality: "USA (Colorado: Salida). . . ." *Pediocactus Hermannii* W. T. Marshall. *Pediocactus Simpsonii* (Engelm.) Britton & Rose var. *Hermannii* W. T. Marshall. *Pediocactus Simpsonii* var. *Hermannii* Wiegand & Backeberg. Type locality: Hatch, Garfield County, Utah.

2. *Pediocactus Bradyi*
   L. Benson

Stems solitary or rarely two, subglobose to obovoid, 1½ to 2½ inches long, 1 to 2 inches in diameter; areoles elliptic, densely white- or yellow-villous; spines dense on the joint, obscuring the stem; central spines none or in one collection 1 or 2 per areole, straight, darker than the radials, about ⅙ inch long; radial spines white or yellowish-tan, glabrous, smooth, cartilaginous, about 14 or 15 per areole, spreading and nearly pectinate, almost straight but the tips curving slightly downward, about equal, ⅛ to ¼ inch in diameter, tapering gradually from bulbous bases, nearly circular in cross section; flowers ⅝ to 1¼ inches in diameter, ⅝ to ⅞ inch long; petaloid perianth parts pale straw-yellow, oblan-

ceolate, up to 1/2 or 5/8 inch long, 1/8 to 3/16 inch broad, acute and apically minutely denticulate, marginally entire; fruit green, drying and tardily turning brownish, smooth except for irregularity at the positions of the areoles, without surface appendages, broadly top-shaped, 1/4 inch long, 5/16 inch in diameter, basally constricted into a short stalk, the top slightly convex, thin and membranous; seeds black, with minute beadlike papillae, but these on larger irregular projections, 1/12 inch long, 1/11 inch broad.

DISTRIBUTION: Colorado Plateau at about 4,000 feet elevation. Arizona in Coconino County near the Marble Canyon of the Colorado River. Navajoan Desert. The species occurs at least sparingly over an area about fifteen miles long.

SYNONYMY: Pediocactus Bradyi L. Benson. *Toumeya Bradyi* W. H. Earle. Type locality: near the Marble Canyon of the Colorado River, Coconino County, Arizona.

3. *Pediocactus Paradinei*
   B. W. Benson

Stems solitary, 1 to 2 inches high, 1 to 2 or reportedly 3¼ inches in diameter; areoles circular; central spines hairlike, dense or sparse in some young individuals, white to pale gray, becoming straw- or cream-color in age, apically sometimes darker, 4 to 6 per areole, the centrals and approximately 20 radials not clearly distinguished, all flexible, straight or curving irregularly, the longer ones 1 to 1 5/16 inches long, basally about 1/96 to 1/72 inch in diameter, elliptic to nearly circular in cross section; flower ¾ to 1 inch in diameter; petaloid perianth parts white with pink midribs, oblanceolate-obovate, up to ¾ inch long, up to ¼ inch broad, rounded and mucronulate, minutely denticulate; fruit greenish-yellow or becoming tan, smooth except for the veins and bare except sometimes for minute subapical scales, nearly cylindroid but enlarged upward, 5/16 to 7/16 inch long, 3/16 to 1/4 inch in diameter; seeds nearly black, tessellate-tuberculate, 1/16 inch long, 3/32 inch broad.

DISTRIBUTION: Gravelly soils of alluvial fans and flats in the desert or grassland at

Fig. 8.1. *Pediocactus Bradyi.* Some of the plants are in fruit. (From Lyman Benson, Cactus and Succulent Journal. Used with permission.)

Fig. 8.2. *Pediocactus Sileri* in flower. (Photo by Robert H. Peebles.)

5,000 to 6,000 feet elevation. Navajoan Desert. Northern Arizona from northeastern Mohave County to the vicinity of Houserock Valley, Coconino County.

SYNONYMY: Pediocactus Paradinei B. W. Benson. *Pilocanthus Paradinei* B. W. Benson & Backeberg. Type locality: Houserock Valley, Coconino County, Arizona.

4. *Pediocactus Sileri*
   (Engelm.) L. Benson

Stems solitary, depressed-ovoid or sometimes ovoid, 2 to 4 or 5 inches long, 2 to 3 or 4 inches in diameter; areoles circular; spines dense; central spines wholly or partly brownish-black, becoming pale gray or nearly white in age, 3 to 7 per areole, standing at right angles to the stem, straight or slightly curving at the tips, 3/4 to 1 1/8 inches long, basally 1/32 to 1/24 inch in diameter, nearly circular in cross section, each tapering from base to apex; flower about 1 inch in diameter; petaloid perianth parts yellowish or straw-color with maroon veins, obovate-oblanceolate, 5/8 to 3/4 inch long, 3/16 to 1/4 inch broad, apically rounded, minutely denticulate; fruit greenish-yellow, with several long-fimbriate scales above, nearly cylindroid but enlarged upward, 1/2 to 5/8 inch long, 1/4 to 3/8 inch in diameter; seeds gray, finely tessellate-tuberculate, obliquely obovate, 1/8 inch long, 3/16 inch broad.

DISTRIBUTION: Hills and flats in the desert at 4,700 to 5,000 feet elevation. Navajoan Desert; Mojavean Desert. Northern

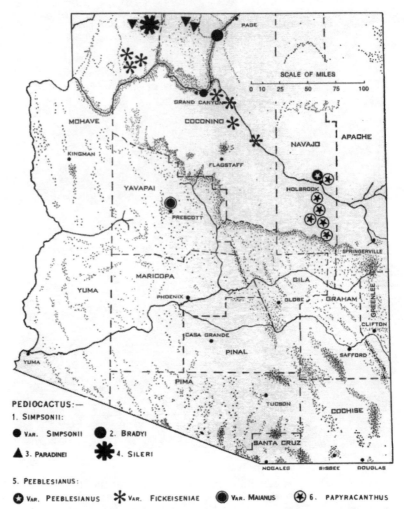

Fig. 8.3. The documented distribution of the Arizona species of *Pedio-cactus*.

Arizona in the vicinity of Pipe Spring and Fredonia; reported from adjacent Utah.

SYNONYMY: Echinocactus Sileri Engelm. *Utahia Sileri* Britt. & Rose. *Pediocactus Sileri* L. Benson. Type locality: "Cottonwood Springs and Pipe Springs, southern Utah [Arizona]."

5. *Pediocactus Peeblesianus*
   (Croizat) L. Benson

Stem(s) obscure, solitary or rarely clustered, somewhat glaucous, obovoid, globose, ovoid-cylindroid, or depressed-globose, often with only the summit protruding above ground, largely retracted into the soil in dry weather, up to 1 to 1½ or 2½ inches long, up to ⅝ to 1½ or 1¾ inches in diameter; areoles circular; spines nearly covering the surface of the stem but not obscuring it; central spines in two varieties 0, in the other 1 per areole, ashy white to pale gray, flexible, turned upward, curving slightly upward, remarkably variable in size from plant to plant, the longer ones ¼ to ¾ or 1⅜ inches long, 1/48 to 1/24 inch in basal diameter, circular to elliptic in cross section; surface of the spine and the tissues beneath remarkably spongy-fibrous; radial spines 3 to 7 per areole, similar to the central but usually much smaller, described further in Table 17; flower about ⅝ to 1 inch in diameter; petaloid perianth parts yellow to yellow-green, sometimes pale or reportedly sometimes white, usually with a middle band of green or pale pink, lanceolate,

Fig. 8.4. *Pediocactus Peeblesianus* var. *Fickeiseniae*. One plant, *below*, is in flower. Note the variability in central spines from one individual to another in a collection from a single area about 100 feet in diameter. (From Lyman Benson, Cactus and Succulent Journal. Used with permission.)

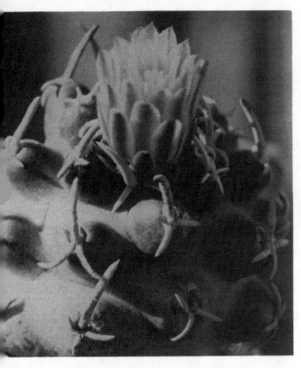

Fig. 8.5. *Pediocactus Peeblesianus*. Mature plant less than 1 inch in diameter, in flower. (Photograph by Robert H. Peebles. From Thomas H. Kearney and Robert H. Peebles, *Arizona Flora*. University of California Press, Berkeley. 1951 and 1960. Used with permission.)

¼ to ⅜ inch long, about ⅛ inch broad, acute to subacute, mucronulate to apiculate, minutely serrulate; fruit greenish, changing to tan, without surface appendages or with 1 or a few scales on the upper portion, subcylindroid, 1/4 to 3/8 inch long, 3/16 to 5/16 inch in diameter; seed black to dark grayish-brown or dark gray, 1/16 to 1/12 inch long, 1/10 inch broad.

5A. Var. **Fickeiseniae**
L. Benson, **var. nov.** (p. 24)
(Table 17)

DISTRIBUTION: Exposed layers of rock on the margins of canyons or hills in the desert at about 4,000 to 5,000 feet elevation. Navajoan Desert and the Great Plains Grassland. Northern Arizona from northeastern Mohave County to the vicinity of the Colorado and Little Colorado rivers in the Grand Canyon region and southeastward in Coconino County. The known range is about 125 airline miles.

This taxon was discovered first by Miss Maia Cowper and shown to Mr. and Mrs. Denis Cowper of Belen, New Mexico, in May, 1956. Specimens were shown by Mrs. Jane Cowper to the late Mr. W. Taylor Marshall, Director of the Desert Botanical Garden at Phoenix, Arizona, and this is the basis for a published note, Saguaroland Bulletin 10:76. cover illustration, p. 73. 1956.

"Mrs. Cowper now reports the discovery of another colony of the plants [interpreted as *Toumeya Peeblesiana* and not considered as even varietally segregated] near Cameron, Coconino County, and about one hundred miles north and west of the first location [Holbrook area]. Here the plants are much larger and healthier and resemble the one plant collected north of Joseph City."

5B. Var. **Maianus**
L. Benson, **var. nov.** (p. 24)
(Table 17)

DISTRIBUTION: Arizona at or near Prescott, Yavapai County.

5C. Var. *Peeblesianus*
(Table 17)

DISTRIBUTION: Hills in the desert at about 5,100 to 5,200 feet elevation. Navajoan Desert. Northern Arizona in Navajo County from near Joseph City to the Marcou Mesa region northwest of Holbrook.

SYNONYMY: Navajoa Peeblesiana Croizat. *Toumeya Peeblesiana* W. T. Marshall. *Echinocactus Peeblesianus* L. Benson. *Pediocactus Peeblesianus* L. Benson. Type locality: vicinity of Holbrook, Navajo County, Arizona.

6. *Pediocactus papyracanthus*
(Engelm.) L. Benson
Grama grass cactus

Stem solitary, elongate, cylindroidal or obconical-cylindroidal, 1 to 3 inches long, ½ to ¾ inch in diameter; areoles circular; spines dense; central spines whitish or pale brown, changing to gray, flexible, 1 or sometimes 2 to 4 per areole, the upper ones smaller, curving upward, the mass of centrals overarching the stem-apex, usually ¾ to 1¼ inches long, basally up to 1/20 inch broad, strongly flattened, involute and the midrib evident on the ventral side, it and the margins puberulent; radial spines ashy white or pale gray, flexible, 6 to 8 per areole, spreading parallel to the stem surface, straight, about ⅛ inch long, up to nearly 1/48 inch broad, flat, very thin; flower ¾ to 1 inch in diameter; petaloid perianth parts white with brownish midribs, oblanceolate, about 3/4 inch long, up to 3/16

Fig. 8.6. Grama grass cactus, *Pediocactus papyracanthus.* The papery spines resemble the leaves of the blue grama grass, *Bouteloua gracilis,* in tufts of which the plant often grows. (Photograph by Robert H. Peebles.)

TABLE 17. CHARACTERS OF THE VARIETIES OF PEDIOCACTUS PEEBLESIANUS

	A. Var. **Fickeiseniae**	B. Var. **Maianus**	C. Var. **Peeblesianus**
Relative size	Larger in all parts.	Larger in all parts.	Smaller in all parts.
Stem	Unbranched or with 2 to 4 branches, up to 1 or 1½ or 2½ inches long, 1 to 1½ inches in diameter.	Unbranched, about 2½ inches long, 1½ inches in diameter.	Unbranched, up to 1 inch long, ⅝ to ¾ or 1 inch in diameter.
Central spine	1, erect and prominent (or small or absent in young plants), clearly differentiated from the radials, highly variable.	None.	None; the upper radial spine often longer than the others and up to ¼ or even 5/16 inch long.
Radial spines	Usually 6 but sometimes 7, straight, spreading irregularly, of varying sizes, ⅛ to ¼ inch long, 1/96 to 1/48 inch in diameter.	6, the three lower stout; about ½ inch long, 1/24 inch in diameter, the lowest one curving strongly; the upper as long but more slender, the 2 upper lateral ones much smaller.	Usually 4 but in some areoles sometimes 3 or 5, recurving, with the appearance of a cross, the lower ones usually ⅛ to 3/16 or ¼ inch long, 1/72 to 1/24 inch in diameter.
Geographical distribution	Arizona from northeastern Mohave County to the Grand Canyon region and the vicinity of the Little Colorado River, Coconino County.	Arizona near Prescott, Yavapai County.	Arizona near Joseph City and Holbrook, Navajo County.

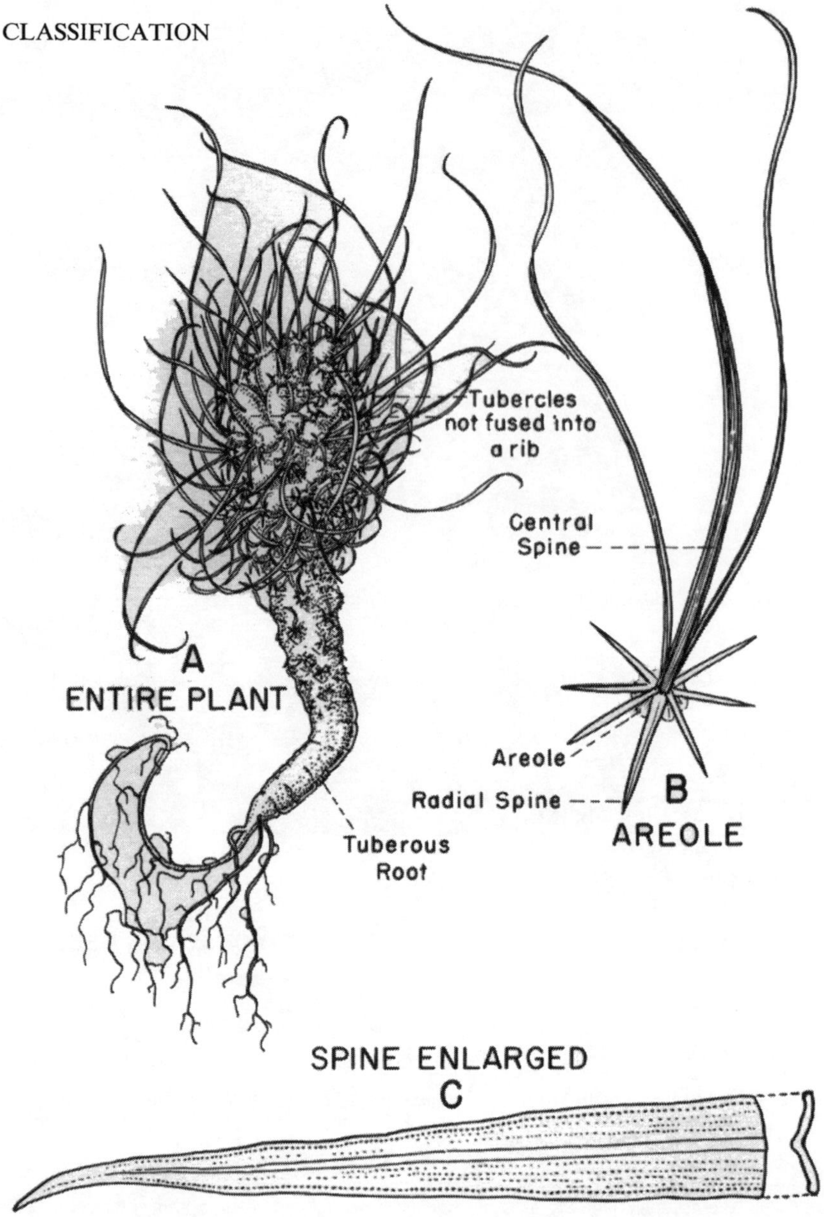

Fig. 8.7. Grama grass cactus, *Pediocactus papyracanthus*

inch broad, sharply acute, mucronate, marginally entire; fruit green, changing to tan, with a few or no persistent scales, subglobose, about 3/16 to 3/4 inch long, up to 3/16 in diameter; seeds black, finely papillate-tessellate, about 1/10 inch long, 1/8 inch broad.

DISTRIBUTION: Open flats in grasslands and woodlands at 5,000 to 7,300 feet elevation. Juniper-Pinyon Woodland and Great Plains Grassland. Arizona in southern Navajo County. Western New Mexico. Inconspicuous and irregular in occurrence and commonly overlooked. The plants grow in or near fairy rings of blue grama grass, and they are unnoticed because the spines resemble the dried leaves of the grass.

SYNONYMY: Mammillaria papyracantha Engelm. *Echinocactus papyracanthus* Engelm. *Toumeya papyracantha* Britton & Rose. *Pediocactus papyracanthus* L. Benson. Type locality: near Santa Fe, New Mexico.

9. EPITHELANTHA
Button Cactus

Stems *either* unbranched *or* branching and in clumps, irregularly ovoid-cylindroid to cylindroid, 1 to 2½ inches long, of about equal diameter; ribs none, the tubercles separate. Leaves not discernible on mature tubercles. Areoles nearly circular. Spines smooth, white or ashy-gray, those in the upper part of the areole longer, making a tuft in the depression at the apex of the stem, ultimately breaking at the middles and the terminal tuft of spines giving way to the area of "shorter spines" along the sides of the stem, 20-100 per areole, in 2 to 5 series, straight, ⅛ to ¼ inch long, basally about 1/240 inch in diameter, needle-like. Flowers and fruits on the new growth of the current season, therefore near the apex of the stem or branch, each at the apex of a tubercle in a felted area against the spine-bearing part of the areole and merging with it, the scar persisting many years and circular to semicircular. Flower ⅛ to ½ inch in diameter; superior floral tube funnelform, the color of the perianth but paler. Fruits red, or some small and colorless and with only 1 to 5 (instead of 5 or 6 to 11) seeds, fleshy, without surface appendages, widening upward, ⅛ to ½ inch in diameter, not splitting open. Seeds black, impressed-reticulate and sometimes also papillate, broader than long (length being hilum to opposite side), 1/24 to 1/16 inch in greatest dimension; hilum appearing "lateral."

DISTRIBUTION: A few species occurring from Arizona to Texas and northern Mexico. Two species in the United States. One in Arizona.

1. *Epithelantha micromeris*
(Engelm.) Weber

Stems solitary or sometimes in small clumps up to 2½ inches high and 6 inches in diameter, obscured by the white or pale gray spines, irregularly ovoid-cylindroid, with a slight apical depression, up to 1½ or 2½ inches long, 1 to 2½ inches in diameter; tubercles conic-cylindroid, in 20 to 35 rows; spines numerous, obscuring the stem, white or pale ashy-gray, in 2 series of about 20 (or with a third series) per areole, accompanied by white woolly hairs, up to ¼ inch long, basally about 1/240 inch in diameter, needle-like, with numerous forward-directed minute barbs; flower small, usually obscured by the long spines near the apex of the stem, about ⅛ to 3/16 inch in diameter; petaloid perianth parts pale pink, approaching deltoid but with the upper sides somewhat curving, 1/24 inch long, 1/24 inch broad, acute, entire; fruits red (or some of them small and colorless and with only 1 to 3 seeds, the seeds usually 4 to 11) fleshy, with no surface appendages, enlarged upward, the lower portion not producing seeds, 1/8 to 1/2 inch long, 1/16 to 1/4 inch in diameter, the superior floral cup deciduous and the apical cup inconspicuous; seed black, the surface minutely reticulate-impressed, 1/24 inch long, 1/16 inch broad.

DISTRIBUTION: Rocky hills and ridges in the desert and in grassland at 3,400 to 5,000 feet elevation. Chihuahuan Desert and the adjacent grasslands. Arizona, rare in Santa Cruz and Cochise counties (collected first by Pringle in 1884). Southern New Mexico; western Texas. Northern Mexico.

SYNONYMY: Mammillaria micromeris Engelm. *Cactus micromeris* Kuntze. *Epithelantha micromeris* Weber. Type locality: San Felipe Creek to the Pecos River, Texas.

10. NEOLLOYDIA

Stems branched or unbranched, ovoid or cylindroid, 2 to 6 or 15 inches long, 1 to 3 or 5 inches in diameter; tubercles separate or nearly separate. Leaves not discernible on the mature stem. Areoles nearly circular to elliptic. Spines smooth, ranging from black to dark brown or tan, chalky blue, straw-color, purplish, or pink, sometimes dark-

tipped, black or gray in age; central spines 1 to 8 per areole, straight, ½ to 1¾ inches long, basally 1/60 to 1/16 inch in diameter, needlelike or flattened; radial spines white or resembling the centrals but usually smaller and of lighter color, 3 to 32 per areole, straight, ¼ to 1¼ inches long, basally 1/120 to 1/16 inch in diameter or breadth. Flowers and fruits on the growth of the current season near the apex of the stem or branch, each on

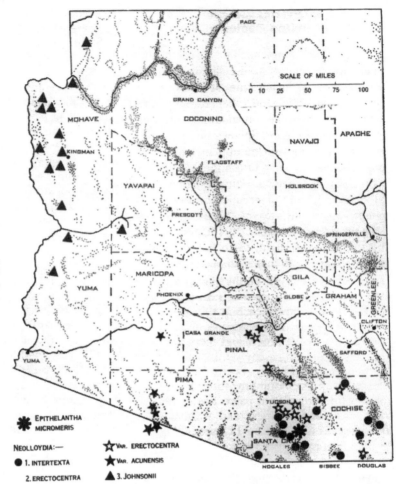

Fig. 10.1. The documented distribution of *Epithelantha micromeris* and of the Arizonan species of *Neolloydia*.

the upper side of a tubercle in a felted area distant from the spine-bearing portion of the areole and connected by an isthmus running the length of the tubercle. the area persisting and forming an elongate and narrow scar. Flower 1 to 3 inches in diameter; superior floral tube funnelform to obconical, green and pink to purple. Fruit dry, with a few or up to 20 broad, membranous scales, ellipsoid to short-cylindroid, ¼ to ½ inch long, about ¼ to ⅜ inch in diameter, opening basally or by 1 to 3 lengthwise slits. Seeds black, reticulate, reticulate-papillate, or papillate, broader than long or longer than broad (length being hilum to opposite side), 1/16 to 1/10 inch in greatest dimension; hilum either obviously basal or appearing "lateral."

DISTRIBUTION: Perhaps twenty or more

species occurring from Arizona to Texas and southward to northern Sonora and San Luis Potosí in Mexico. Seven species in California, Arizona, New Mexico, and Texas. Three in Arizona.

LIST OF SPECIES AND VARIETIES

The combinations under *Thelocactus* were based upon basionyms published by Engelmann or, for *Thelocactus Johnsonii,* Parry ex Engelmann under *Echinocactus.* The combinations under *Thelocactus* appeared in Kelsey & Dayton, Standardized Plant Names, Ed. 2. p. 78. 1942, with faint references to the basionyms. The work on the "cactus genera" was credited to the advice of Elzada U. Clover, **W.**

Taylor Marshall, and R. W. Poindexter. In view of later publication, as *nomina nuda,* of *Thelocactus intertextus* and *Thelocactus erectocentra* by W. T. Marshall, Cactus & Succ. Jour. Gt. Brit. 9:28. 1947, and, with indirect reference to the basionym through *Ferocactus Johnsonii* Britton & Rose, of *Thelocactus Johnsonii* by the same author, *ibid.,* the combinations are attributed here to Marshall.

KEY TO THE SPECIES

1. Fruit dehiscent along a diagonal line of abscission at the base; central spines basally 1/60 to 1/44 inch in diameter; radial spines 1/120 to 1/48 inch in diameter; lower central spine much shorter than the radials.
   1. *Neolloydia intertexta,* page 191
1. Fruit dehiscent along 1 to 3 vertical slits; central spines basally 1/24 to 1/20 or 1/16 inch in diameter; radial spines 1/48 to 1/36 inch in diameter.
   2. Flowers pink or pale pink; central spines 1 to 4, *either* (1) solitary *or* (2) dissimilar in length or in bending or curvature, the upper 1 to 3 spreading more or less like the radials, the principal one either protruding or turned upward, much longer than the radials.
      2. *Neolloydia erectocentra,* page 191
   2. Flowers magenta to yellow; central spines 4 to 9, all nearly alike, spreading in all directions.   3. *Neolloydia Johnsonii,* page 192

1. Neolloydia intertexta
(Engelm.) L. Benson
comb. nov. (p. 24)

Stems solitary, cylindroid at maturity, 2 to 3 or 6 inches long, 1½ to 3 inches in diameter; stem indented deeply and usually sharply between tubercles; spines dense, covering and appressed against the stem, each forming a low arc; central spines pinkish or all but the tips grayish straw-color, usually 4 per areole, the three upper pointing more or less upward and appearing to be radials, ⅝ to ⅞ inch long, basally up to 1/44 inch in diameter, needle-like, the lower (obviously central) one similar to the upper or at right angles to the stem, 1/24 to 1/8 inch long, elongate-conical; radial spines similar to the upper centrals, 13 to 20 or (reportedly) 25 per areole, spreading, curving slightly, the longer ones ⅜ to ⅝ or ⅘ inch long, basally about 1/48 inch in diameter, elliptic in cross section; flower 1 to 1¼ inches in diameter, ¾ to 1¼ inches long; petaloid perianth parts pink or pale pinkish-white on the margins, each one with a broad pink median band, lanceolate, up to ¾ inch long, up to ⅛ inch broad, acute or sharply acute to cuspidate, entire or essentially so;

fruit green, changing to tan or brown, with a few short, broad, membranous scales, ellipsoid-cylindroid, about ½ inch long, about ¼ inch in diameter, with the apical cup covered by the persistent superior floral tube; seed black, minutely and regularly papillate-reticulate except near the hilum, 1/16 inch long, 1/2 inch broad.

DISTRIBUTION: Limestone hills in grassland at 4,000 to 5,000 feet elevation. Desert Grassland and upper edge of the Chihuahuan Desert. Arizona from southern Pima County to Santa Cruz and Cochise counties. Eastward in New Mexico to Bernalillo and Doña Ana counties, and in Texas to Jeff Davis and Brewster counties. Mexico in northern Sonora and Chihuahua.

SYNONYMY: Echinocactus intertextus Engelm. *Echinomastus intertextus* Britton & Rose. *Thelocactus intertextus* W. T. Marshall. Type locality: El Paso to Limpia Creek, Texas. *Cereus pectinatus* Engelm. var. *centralis* Coulter. *Echinocereus pectinatus* var. *centralis* K. Schum. *Echinocereus centralis* Rose. *Echinomastus centralis* Y. Ito. Type locality: near Fort Huachuca, Arizona.

2. Neolloydia erectocentra
(Coulter) L. Benson
comb. nov. (p. 24)

Stems solitary, ovoid, or somewhat cylindroid, 3 to 6 or 15 inches long, 3 to 4 or 5 inches in diameter; spines obscuring the surface of the stem; central spines dark-tipped, reddish, pink, or purplish or the lower halves straw-color, 1 to 4 per areole, the upper turned upward, the lower (principal) one spreading or turned somewhat upward, straight or slightly curving, ¾ to 1⅜ inch long, basally about 1/24 inch in diameter, circular in cross section; radial spines similar to the central or straw-color and of about half the diameter, 11 to 15 per areole, appressed, the lateral pectinate, straight, 1/2 or sometimes 5/8 inch long, basally 1/48 to 1/36 inch in diameter, nearly circular in cross section; flower 1½ to 1¾ or 2 inches in diameter; petaloid perianth parts pink or rarely white, spathulate-oblanceolate, up to 1 1/4 or 1 1/2 inches long, up to 7/16 inch broad, cuspidate, entire; fruit green, drying to tan, with several membranous scales and these minutely and sharply denticulate, subcylindroid, about 3/8 inch long, about 5/16 inch in diameter, opening along a dorsal slit or

Fig. 10.2. *Neolloydia erec-tocentra*. Spine cluster.

along two irregular lines; seeds minutely and regularly papillate except near the hilum, 1/16 inch long, 1/12 inch broad.

2A. Var. *erectocentra*

Stems 4 to 6 or 15 inches long, up to 3 or 5 inches in diameter; tubercles vertically about 1/4 inch long; central spines 1 or 2, 1/2 to 7/8 inch long, basally about 1/48 inch in diameter; radial spines 1/2 or 5/8 inch long, basally about 1/48 inch in diameter; flower 1½ to 1¾ inches in diameter, about 1½ inches long; petaloid perianth parts pink.

DISTRIBUTION: Alluvial fans and hills in the desert and grassland at about 3,000 to 4,300 feet elevation. Upper Arizona Desert and Desert Grassland. Southeastern Arizona from southeastern Pima County to western Cochise County.

SYNONYMY: Echinocactus erectocentrus Coulter. *Echinocactus horripilus* Lemaire var. *erectocentrus* Weber. *Echinomastus erectocentrus* Britton & Rose. *Thelocactus erectocentrus* W. T. Marshall. Type locality: near Benson, Cochise County, Arizona. *Mammillaria Childsii* Blanc. According to Britton & Rose, Cactaceae 3:175. 1922, the description and figure indicate this species. *Echinocactus Krausei* Hildmann. *Thelocactus Krausei* Kelsey & Dayton, *nom. nud.*, based upon *Echinomastus Krausei*, then a *nom. nud.* Type locality: Dragoon Summit, Cochise County, Arizona.

2B. Var. **acunensis**
(W. T. Marshall) L. Benson, **comb. nov.** (p. 25)

Stems up to 9 inches long and 4 inches in diameter; tubercles about ½ inch long vertically; central spines 3 or 4, up to 1 or 1⅜ inch long, basally about 1/24 inch in diameter; radial spines about 1 inch long, up to 1/36 inch in diameter; flower about 2 inches in diameter; petaloid perianth parts coral pink to mallow.

DISTRIBUTION: Hills and flats in the desert at 1,300 to 2,000 feet elevation. Arizona from western Pima County to the Sand

Tank Mountains in Maricopa County and to Pinal County. Mexico in adjacent Sonora.

SYNONYMY: Echinomastus acunensis W. T. Marshall. Type locality: eastern Organ Pipe Cactus National Monument, Pima County, Arizona.

3. **Neolloydia Johnsonii**
(Parry) L. Benson
comb. nov. (p. 25)

Solitary or rarely branching following injury to the terminal bud; stems ovoid- to ellipsoid-cylindroid, mostly 4 to 6 or 10 inches long, 2 to 4 inches in diameter; spines obscuring the stem; central spines pink to red, becoming blackened in age, 4 to 8 per areole, spreading widely, straight or slightly curved, 1¼ to 1½ inches long, basally 1/24 to 1/20 or 1/16 inch in diameter, nearly circular in cross section; radial spines similar but smaller and lighter in color, about 9 to 10 per areole, up to ½ to ¾ or 1 inch long, basally about 1/48 to 1/36 inch in diameter, circular to elliptic in cross section; flower 2 to 3 inches in diameter; petaloid perianth parts magenta to pink or greenish-yellow, broadly oblanceolate, about 1¼ inches long, ⅜ inch broad, acute to mucronulate, entire; fruit green, drying to tan, with several scarious-margined fringed scales, ellipsoid-cylindroid, about ½ inch long, ⅜ inch in diameter, splitting along the dorsal (back) side; seeds minutely papillate except near the hilum, 1/12 inch long, 1/10 inch broad.

DISTRIBUTION: Hills and alluvial fans in the desert at 1,700 to 4,000 feet elevation. Mojavean Desert. Arizona in Mohave, western Yavapai, and northern Yuma counties. California in the Death Valley Region; Nevada in Clark County; Utah in Washington County.

SYNONYMY: Echinocactus Johnsonii Parry. *Ferocactus Johnsonii* Britton & Rose. *Echinomastus Johnsonii* Baxter. *Thelocactus Johnsonii* W .T. Marshall. Type locality: about St. George, Washington County, Utah. *Echinocactus Johnsonii* Parry var. *octocentrus* Coulter. Type locality: Resting Springs Mountains, Inyo County, California. *Echinocactus Johnsonii* Parry var. *lutescens* Parish. *Echinomastus Johnsonii* (Parry) Britton & Rose var. *lutescens* Parrish [Parish] ex Glade, *nomen nudum.* Type locality: hills at Searchlight, Clark County, Nevada. *Echinomastus arizonicus* Hester. Type locality: Butler Valley, Yuma County, Arizona.

Fig. 10.3. *Neolloydia. Above, Neolloydia Johnsonii,*
in flower. *Below, Neolloydia erectocentra* var.
acunensis. (Photograph, *above,* by Evelyn L.
Benson.)

11. CORYPHANTHA

Stems solitary or branching, in some species forming mounds of 200 or more, at maturity subglobose to cylindroid, 1 to 4 or 6 inches long, 1 to 3 inches in diameter; ribs none, the tubercles separate. Leaves not discernible. Areoles circular to elliptic. Spines smooth, white, gray, pink, yellow, brown or black; central spines 1 to 10 per areole or more and grading into the radial, straight, curved, hooked, or twisted, ⅛ to 1 or 2 inches long, basally 1/120 to 1/24 inch in diameter or width, narrowly to broadly elliptic in cross section; radial spines commonly of the same color as or lighter than the centrals, 5 to 40 per areole, usually straight, ⅛ to 1 inch long, basally 1/240 to 1/48 inch in diameter, needlelike. Flowers and fruits on the new growth of the current season, in mature stems located below the apex of the stem, each one at the base of the upper side of a tubercle distant from the spine-bearing part of the areole and connected with it by a narrow isthmus, the scar persisting for many years, elongate and forming a felted groove; on immature stems the flower-bearing area as high as mid-level on the tubercle. Flower ½ to 2 inches in diameter; superior floral tube funnelform, green or tinged with pink, magenta, reddish-purple, yellow-green, or yellow. Fruit fleshy, green or red, thin-walled, without surface appendages, usually ellipsoid, club-shaped, or cylindroid, 1/4 to 2 1/4 inches long, 1/12 to 3/4 inch in diameter, with the floral tube deciduous, indehiscent. Seeds tan, brown, or black and smooth and shining or punctate or reticulate, *usually* broader than long (length being hilum to opposite side), 1/24 to 1/12 inch in greatest dimension; hilum usually appearing "lateral," sometimes obviously basal or "oblique."

DISTRIBUTION: Twenty or thirty species from southern Alberta to central Mexico. Fourteen species from Oregon and Alberta to California and Texas. Six in Arizona.

LIST OF SPECIES AND VARIETIES

KEY TO THE SPECIES

1. Tubercles not necessarily persistent on the lower part of the stem after the spine-clusters fall, not becoming hard, the old spineless tubercles usually inconspicuous.
 2. Fruit green at maturity, commonly cylindroid, club-shaped, or ellipsoid; seeds usually brown, broader than long, the hilum appearing "lateral"; central spines usually present and different from the radials.
 3. Flowers at the apex of the stem at the upper edge of a single series of new tubercles just developing in the current blooming period (fruits of plants blooming more than once a year relatively lower on the stem).
 4. Tubercles 1 to 1½ inches long, conspicuous; bases of the central spines 1/24 to 1/16 inch in diameter, straw-yellow with dark red tips.
1. *Coryphantha Scheeri,* page 195
 4. Tubercles not more than ½ or rarely ¾ inch long; bases of the central spines 1/60 to 1/44 inch in diameter.
2. *Coryphantha vivipara,* page 196
 3. Flowers appearing (at the same time) on several turns of a spiral near the apex of the plant, some from maturing tubercles more or less at the side of the stem; principal or commonly the only central spine (there being occasionally 2) curving downward in a low arc.
3. *Coryphantha recurvata,* page 202
 2. Fruit red, globular; seeds black, slightly longer (hilum to opposite point) than broad, the hilum obviously basal; central spine(s) solitary or lacking or rarely 2, except in position little different from the radials, straight.
5. *Coryphantha missouriensis,* page 204
1. Tubercles persistent on the lower part of the stem after the spine clusters fall, becoming hard, the old spineless tubercles conspicuous in older plants; seed broader than long (hilum to opposite point), the hilum appearing "lateral"; seed coat brown; stems usually branching; fruit usually bright red (though sometimes green) at maturity.
4. *Coryphantha strobiliformis,* page 202

1. Coryphantha Scheeri
(Kuntze) L. Benson
comb. nov. (p. 25)

Stems solitary or sometimes in clusters, ellipsoid, 4 to 7 inches long, 3 to 4 inches in diameter; spines dense; central spines straw-color with dark red tips or pink, 1 to 4 or 5 per areole, spreading, 1½ inches long, basally up to 1/24 or 1/16 inch in diameter, nearly circular in cross section; radial spines like the central, 1 to 4 or 5 per areole, spreading irregularly, straight, curving a little, or hooked, up to 1⅛ inches long, basally 1/32 inch in diameter, elliptic in cross section; flower 2 to 3 inches in diameter; petaloid perianth parts yellow or with red streaks, narrowly oblanceolate, 1¼ inches long, up to ⅜ inch broad, from slightly acuminate to acute, entire or with a few sharp teeth; fruit green, ellipsoid, 1¼ to 2¼ inches long, ½ to ¾ inch in diameter; seeds brown or perhaps becoming black, finely reticulate, ovate-acute, 1/12 inch long, ⅛ inch broad.

1A. Var. valida
(Engelm.) L. Benson, **comb. nov.** (p. 25)

Stems solitary; areoles at first covered densely with wool, which disappears later; central spines 1 to 4 or 5, the primary one up to 1½ inches long, 1/24 inch in diameter, not apically strongly curved or hooked; radial spines 9 or 12 to 16, up to 1¼ inches long; sepaloid perianth parts ciliate (fringed with hairs); petals yellow; floral tube relatively broad.

DISTRIBUTION: Deep soils of flats and bottomlands in deserts and grasslands at about 4,000 feet elevation. Desert Grassland and Chihuahuan Desert. Southeastern Arizona near Nogales in Santa Cruz County, in southern Pima County, and in San Simon Valley, Cochise County. Southern New Mexico as far east as Doña Ana County; Texas from El Paso County to the Davis Mountains and Alpine. Mexico in northern Chihuahua.

SYNONYMY: Mammillaria Scheeri Mühlenpfordt var. *valida* Engelm., not *Mammillaria valida* J. A. Purpus. Type locality: near El Paso, Texas.

1B. Var. robustispina
(Schott) L. Benson, **comb. nov.** (p. 25)

Stems branching, forming clumps; areoles at first densely woolly, the wool later disap-

Fig. 11.1. *Coryphantha Scheeri* var. *valida.* Plant growing in a Desert Grassland flat which sometimes is flooded. (Photograph by Robert H. Peebles.)

pearing; central spine 1, up to 1⅛ inches long, 1/16 inch or slightly more in diameter, usually curved or hooked and tapering abruptly to the apex; radial spines about 6 in young plants, commonly 10 to 15 at maturity, ¾ to ⅞ inch long, the upper 3 or 4 slender; sepaloid perianth parts fringed with hairs on the lower margins; petaloid perianth parts yellow, salmon, or rarely white; floral tube narrow.

DISTRIBUTION: Alluvial valleys and hillsides in the desert or grassland or woodland at 2,300 to 5,000 feet elevation. Upper Arizona Desert; Desert Grassland; lower Southwestern Oak Woodland. Arizona in Pima County from the Baboquivari Mountains and the Sierrita (south of Tucson) to Santa Cruz County. Mexico in northern Sonora.

SYNONYMY: Mammillaria robustispina Schott. *Cactus robustispinus* Kuntze. *Coryphantha robustispina* Britton & Rose. *Coryphantha Muehlenpfordtii* (Poselger) Britton & Rose var. *robustispina* W. T. Marshall. *Coryphantha Scheeri* (Kuntze) L. Benson var. *robustispina* (Schott) L. Benson. Type

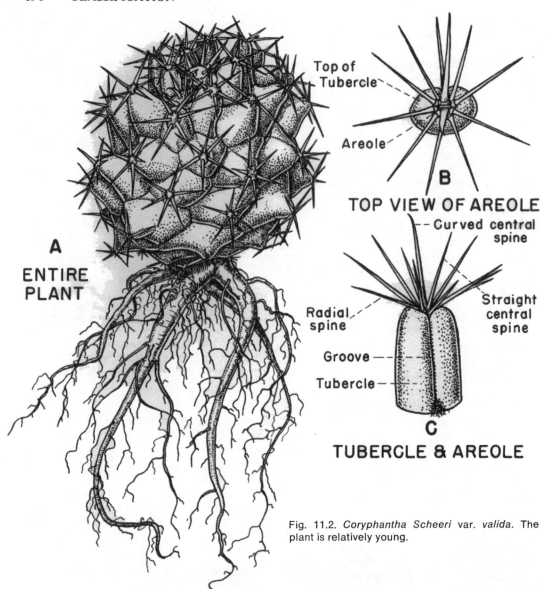

A
ENTIRE
PLANT

Top of
Tubercle

Areole

B

TOP VIEW OF AREOLE

Curved central
spine

Radial
spine

Straight
central
spine

Groove

Tubercle

C

TUBERCLE & AREOLE

Fig. 11.2. *Coryphantha Scheeri* var. *valida*. The plant is relatively young.

locality: south side of the Baboquivari Mountains, Sonora, Mexico. *Mammillaria Brownii* Toumey. *Cactus Brownii* Toumey. Type locality: Baboquivari Mountains, Pima County, Arizona.

2. *Coryphantha vivipara*
   (Nutt.) Britton & Rose

Depressed-globose to ovoid or cylindroid, some varieties forming clumps of 200 or more and up to 1 foot high and 2 feet or more in diameter; stems green, 1½ to 6 inches long, 1½ to 3 inches in diameter; spines dense, partly obscuring the stem; central spines usually white basally but tipped with pink, red, or black for various distances according to the variety, 3 to 10 per areole, spreading, straight, ½ to ¾ or rarely 1 inch long, basally 1/72 to 1/36 inch in diameter, nearly circular in cross section; radial spines white, 12-40 per areole, spreading, straight, ⅜ to ⅝ inch long, basally mostly 1/120 but up to 1/60 inch in diameter, nearly circular in cross section; flowers in some varieties open only an hour or two on one day in the year but in other varieties often several series of

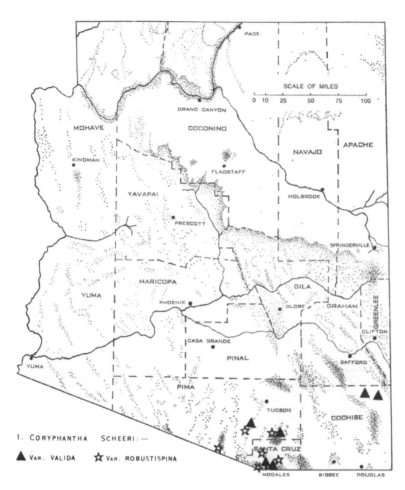

Fig. 11.3. The documented distribution of *Coryphantha Scheeri*, according to its varieties.

flowers produced at different times in a single year; flower 1 to 2 inches in diameter; sepaloid perianth parts fringed; petaloid perianth parts pink, red, lavender, or yellow-green, lanceolate-linear to narrowly lanceolate-attenuate, ½ or ¼ inches long, 1/12 to 3/16 inches broad, mucronate to acuminate or attenuate, entire; fruit green, rarely with a few scales, ellipsoid, ½ to 1 inch long, ⅜ to ⅝ inch in diameter; seeds brown, reticulate, semicircular, but a little broader on one side, 1/22 inch long, 1/16 to 1/12 inch broad; hilum appearing "lateral."

2A. Var. **bisbeeana**
(Orcutt) L. Benson, **comb. nov.** (p. 25)
(Table 18)

DISTRIBUTION: Plains and hills in grassland at 3,000 to 5,200 feet elevation. Desert Grassland. Arizona in eastern Maricopa,

Pinal, Gila, Pima, Santa Cruz, and Cochise counties. Mexico in Sonora adjacent to the area of occurrence in Arizona.

SYNONYMY: Coryphantha bisbeeana Orcutt. *Escobaria bisbeeana* Borg. *Escobaria bisbeeana* W. T. Marshall, *nom. nud. Mammillaria bisbeeana* Orcutt ex Backeberg, *pro syn., nom. nud.,* incorrectly ascribed to Orcutt. Orcutt's intention was clearly to rename the plant described as *Coryphantha aggregata* by Britton and Rose, Cactaceae 3:47. 1922. This was because he adopted Engelmann's reinterpretation of *Mammillaria aggregata* as a synonym of *Echinocereus triglochidiatus* var. *melanacanthus.* Lectotype locality: Benson, Arizona.

2B. Var. *arizonica*
(Engelm.) W. T. Marshall
(Table 18)

DISTRIBUTION: Sandy and rocky soils in the mountains and on high plains in the juniper and yellow pine belts at 4,700 to

TABLE 18. CHARACTERS OF THE VARIETIES OF CORYPHANTHA VIVIPARA

	A. Var. bisbeeana	B. Var. arizonica	C. Var. desertii	D. Var. Alversonii	E. Var. rosea
Stems	Ovoid, solitary or branching, often forming mounds up to 2 feet or more in diameter, each 2 to 3 inches long, 2 to 2½ or 2¾ inches in diameter.	Ovoid, solitary or forming mounds up to 2 feet or more in diameter, each 2 to 4 inches long, 2 to 2½ inches in diameter.	Cylindroid, usually solitary, each 3 to 6 inches long, 2½ to 3½ inches in diameter.	Cylindroid, unbranched but the plant with branching rhizomes under the sand, each 4 to 6 inches long, about 2½ to 3 inches in diameter.	Ovoid-globose, solitary or branching, each 3 to 5 or 7 inches long, 3 to 4 or 6 inches in diameter.
Central spines	5-6, brown or gray with pink or brown tips, ½ to ⅝ inch long, seeming to mingle with the mass of radials.	5-7, red, but white basally, ⅝ to ¾ inch long, except for color seeming to mingle with the mass of radials.	4-6, white, tipped with red, ½ to ¾ inch long, seeming to mingle with the radials, relatively stout.	8-10, white, tipped with dark red or black, ½ to ⅝ inch long, forming a remarkably dense mass with the radials, stout.	10-12, white with red tips, ¾ to 1 inch long, 1/24 inch in diameter, robust.
Radial spines	20-30, white to brown, mostly ⅜ to ½ inch long, of variable diameter within the areole, 1/240 to 1/120 inch in diameter.	20-30, white, ½ to ⅝ inch long, about 1/120 inch in diameter.	12-20, white, ⅜ to ½ inch long, about 1/120 inch in diameter.	About 12-18, white, ½ to ⅝ or ¾ inch long, about 1/72 inch in diameter.	About 12-18, white, ⅝ to ¾ or 1 inch long, about 1/48 inch in diameter.
Flower diameter	1½ to 2 inches.	1½ to 2 inches.	1 to 1¾ inches.	About 1¼ inch.	1¼ to 2 inches.
Petaloid perianth parts	Pink, narrowly lanceolate, the tips acuminate, 1 to 1¼ inches long, about 3/16 inch broad.	Deep pink, narrowly lanceolate, *not* attenuate or acuminate but acute and mucronate, 1 to 1¼ inches long, about 3/16 inch in diameter.	Yellow-green to straw-yellow or pink, lanceolate-linear, *not* attenuate or acuminate but mucronate, about ½ to 1 inch long, about ⅛ inch broad.	Magenta to pink.	Magenta to purplish, lanceolate.
Fruit length	About ¾ inch.	¾ inch.	About 1 inch.	. .	. .
Seed length	1/12 inch.	1/12 inch.	1/16 inch.	. .	1/12 inch.
Geographical distribution	Southern Arizona in Pinal, Pima, Santa Cruz, and Cochise counties. Mexico in adjacent Sonora.	Higher elevations than the other varieties; 4,700 to 7,200 feet. Northern Arizona. Southern Nevada to southwestern Utah and western and north-central New Mexico.	Western edge of Arizona. Clark and Ivanpah Mountains, Mojave Desert in southeastern California, to the Charleston Mountains, Nevada, and southwestern Utah.	Pagumpa, Arizona. California in the Mojave and Colorado Deserts in Riverside County.	Arizona near Peach Springs, Mohave County. California in San Bernardino County; Nevada in Clark and Lincoln counties.

Fig. 11.4. *Coryphantha Scheeri* var. *robusti-spina*. (Photograph by Robert H. Peebles.)

7,200 feet elevation. Juniper-Pinyon Wood-land and Rocky Mountain Montane Forest. Northern and eastern Arizona. Eastern and southern Nevada; southern Utah; western half of New Mexico.

SYNONYMY: Mammillaria arizonica Engelm. *Cactus radiosus* (Engelm.) Coulter var. *arizonicus* Coulter. *Mammillaria radiosa* Engelm. var. *arizonica* Engelm. ex. K. Schum. *Coryphantha arizonica* Britton & Rose. *Mammillaria vivipara* (Nutt.) Haw. var. *arizonica* L. Benson. *Coryphantha vivipara* var. *arizonica* W. T. Marshall. *Escobaria arizonica* F. Buxbaum, not Hester in 1941, *nom. nud.* Type locality: "Arizona."

2C. Var. *desertii*
  (Engelm.) W. T. Marshall
  (Table 18)

DISTRIBUTION: Hills and flats in the desert at 1,000 to 5,400 feet elevation. Mo-javean Desert. Arizona in Mohave County

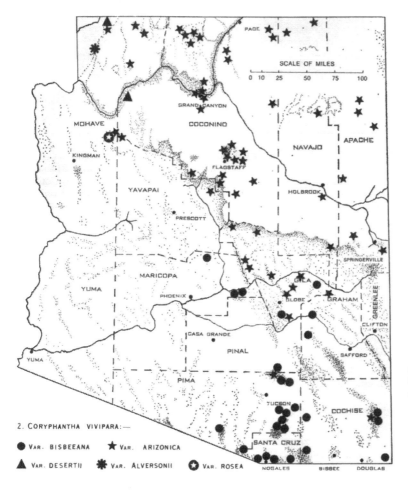

Fig. 11.5. The docu-mented distribution of *Coryphantha vivipara,* ac-cording to its varieties.

Fig. 11.6. *Coryphantha vivipara* var. *arizonica.* Stems forming a mound. (Photograph converted from Kodachrome by Robert C. Frampton Studios, Claremont, California.)

north of Wolf Hole. California in the Mojave Desert in eastern San Bernardino County; Nevada as far northeastward as the Charleston Mountains.

SYNONYMY: Mammillaria desertii Engelm. *Cactus radiosus* (Engelm.) Coulter var. *desertii* Coulter. *Mammillaria radiosa* Engelm. var. *desertii* Engelm. *Coryphantha desertii* Britton & Rose. *Mammillaria vivipara* (Nutt.) Haw. var. *desertii* L. Benson. *Coryphantha vivipara* (Nutt.) Britton & Rose var. *desertii* W. T. Marshall. *Escobaria desertii* F. Buxbaum. Type locality: Ivanpah, San Bernardino County, California. *Mammillaria chlorantha* Engelm. *Cactus radiosus* (Engelm.) Coulter var. *chloranthus* Coulter. *Mammillaria radiosa* Engelm. var. *chlorantha* Engelm. ex K. Schum. *Mammillaria vivipara* (Nutt.) Haw. var. *chlorantha* L. Benson. *Escobaria chlorantha* F. Buxbaum. Type locality: "Southern Utah, east of Saint George."

2D. Var. **Alversonii**
   (Coulter) L. Benson, **comb. nov.** (p. 26)
   (Table 18)

DISTRIBUTION: Sand in the desert at 250 to 4,000 feet elevation. Arizona at Pagumpa. California along the border zone between the Mojave and Colorado deserts in Riverside County and near Bard, Imperial County.

SYNONYMY: Cactus radiosus (Engelm.) Coulter var. *Alversonii* Coulter. *Mammillaria Alversonii* Coulter ex Zeissold, incorrectly ascribed to Coulter. *Mammillaria radiosa* Engelm. var. *Alversonii* K. Schum. *Mammillaria vivipara* (Nutt.) Haw. var. *Alversonii* L. Benson. Type locality: Mojave Desert, California.

2E. Var. **rosea**
   (Clokey) L. Benson, **comb. nov.** (p. 26)
   (Table 18)

DISTRIBUTION: Limestone of gravelly hills in woodlands in the mountains at 5,000 to 9,000 feet elevation. Arizona in the area north of Peach Springs, Mohave County. California in easternmost San Bernardino County; southern Nevada in Clark and Lincoln counties.

SYNONYMY: Coryphantha rosea Clokey. Type locality: Charleston Mountains, Clark County, Nevada. *Coryphantha Alversonii* (Coulter) Orcutt var. *exaltissima* Wiegand & Backeberg, *nom nud. Type* locality: "Kalifornien." No type designated.

Fig. 11.7. Coryphantha. *Above, Coryphantha vivipara* var. *bisbeeana* in flower. *Below, Coryphantha strobiliformis* in flower. (Photo, *above,* by Walter S. Phillips.)

Fig. 11.8. *Coryphantha recurvata.* (Photograph by Robert A. Darrow; converted from Kodachrome by Robert C. Frampton Studios, Claremont, California.)

3. *Coryphantha recurvata* (Engelm.) Britton & Rose

Stems forming clumps, sometimes up to 50, the clumps up to 1 foot high and 1 to 3 feet in diameter; stems green, cylindroid, 4 to 10 inches long, 3-6 inches in diameter; spines obscuring the stem; central spine(s) at first yellow, later gray, the tips red, 1 or rarely 2 per areole, curving downward in a low arc, ⅝ to ¾ inch long, about 1/24 inch in diameter, nearly circular in cross section; radial spines colored like the central, 12 or 15 to 20 per areole, spreading parallel to the stem, each curving slightly in a low arc, ½ to ⅝ inch long, basally about 1/48 inch in diameter, nearly circular in cross section; flower 1 to 1½ inches in diameter; petaloid perianth parts greenish-yellow or reportedly lemon-yellow, oblanceolate, ½ to ⅝ inch long, 1/8 to 3/16 inch broad, mucronate, denticulate-serrulate; fruit green, spheroidal, about 5/16 inch in diameter; seeds not available.

DISTRIBUTION: Alluvial soils of valleys and foothills in grassland and the oak belt at 4,000 to 6,000 feet elevation. Desert Grassland and Southwestern Oak Woodland. Arizona in western Santa Cruz County from Nogales and the Tumacacori Mountains westward. Mexico in adjacent Sonora.

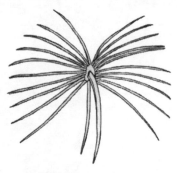

Fig. 11.9. *Coryphantha recurvata.* Spine cluster, showing the downward-curving (recurved) central spine.

SYNONYMY: Mammillaria recurvispina Engelm. in 1856, not De Vriese in 1839. *Mammillaria recurvata* Engelm., *nom. nov. Cactus recurvatus* Kuntze. *Cactus Engelmannii* Kuntze, *nom. nov.* for *recurvispina. Coryphantha recurvata* Britton & Rose. Type locality: Sierra del Pajarito, Pimeria Alta, Sonora, Mexico. *Mammillaria nogalensis* Runge. Type locality: Nogales, Arizona.

4. *Coryphantha strobiliformis* (Poselger) Orcutt.

Stems solitary or forming small clumps several inches in diameter, glaucous-green, cylindroid, 2 to 5 or 8 inches long, 1 to 2 inches in diameter; tubercles, unlike those of

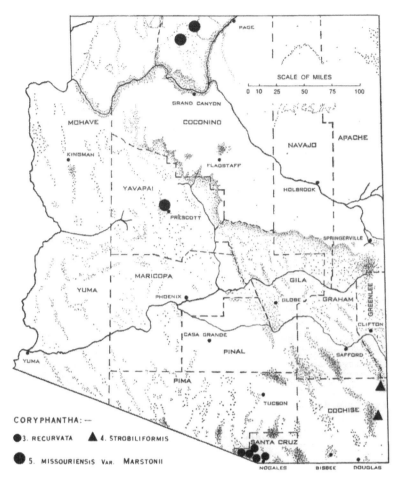

CORYPHANTHA: —

● 3. RECURVATA   ▲ 4. STROBILIFORMIS

● 5. MISSOURIENSIS Var. MARSTONII

Fig. 11.10. The documented distribution of *Coryphantha recurvata*, of *Coryphantha strobiliformis*, and of *Coryphantha missouriensis* var. *Marstonii*.

other Arizona species, becoming spineless, hard, and persistent on the lower part of the stem; spines tending to obscure the stem; central spines straw-yellow, *usually* tipped with pink or pale red, the principal one (central in the areole) surrounded by 5 to 7 smaller ones, straight, ½ to ⅝ inch long, basally 1/60 inch in diameter, nearly circular in cross section; radial spines pale straw color or white or pale gray, 20 to 30 per areole, spreading, straight, ½ inch long, basally 1/120 inch in diameter, nearly circular in cross section; flower ¾ to 1¼ inches in diameter; petaloid perianth parts pink, lanceolate, ⅝ to ¾ inch long, ⅛ to 3/16 inch broad, sharply acute, entire; fruits red or sometimes some of them green, narrowly ellipsoid-cylindroid, ½ to ¾ inch long, 1/4 to 5/16 inch in diameter; seeds brown, punctate, broader than long, 1/32

inch long, 1/24 inch broad; hilum very small, appearing "sublateral."

4A. Var. *strobiliformis*

The most central spine longest, surrounded by 5 to 7 smaller centrals, the principal one up to ½ or ⅝ inch long, up to 1/60 inch in diameter.

DISTRIBUTION: Limestone soils of hills, canyons, and alluvial fans in the desert and in grassland at 2,500 to 5,100 feet elevation. Chihuahuan Desert and Desert Grassland. Arizona in the Peloncillo and Chiricahua Mountains, Cochise County. Southern and central New Mexico; Texas west of the Pecos River. Mexico in Chihuahua.

SYNONYMY: Mammillaria strobiliformis Scheer in 1850, not Engelmann Jan. 13, 1848, not Mühlenpfordt Jan. 15, 1848, therefore a later homonym and illegitimate when published. *Echinocactus strobili-formis* Poselger in 1853, treated as newly published in that year (International Code of Botanical Nomenclature, Article 72), therefore the basionym. *Cactus strobiliformis* Kuntze. *Escobaria strobiliformis* Scheer ex Bödeker. *Coryphantha strobiliformis* Orcutt. Type locality: northern Mexico. *Echinocactus Pottsianus* Poselger, not *Mammillaria Pottsii* Scheer. Type locality: Guerrero, south of the Rio Grande. *Mammillaria tuberculosa* Engelm. *Cactus tuberculosus* Kuntze. *Coryphantha tuberculosa* Orcutt. *Escobaria tuberculosa* Britton & Rose. *Coryphantha tuberculosa* Berger. *Escobaria tuberculosa* Bödeker. Type locality: Flounce Mountains, below El Paso near the Rio Grande, in Chihuahua, Mexico. *Mammillaria strobiliformis* Scheer var. *rufispina* Quehl. *Escobaria tuberculosa* (Engelm.) Britton & Rose var. *rufispina* Borg. No specimen or locality mentioned. *Mammillaria strobiliformis* Scheer var. *pubescens* Quehl. *Escobaria tuberculosa* (Engelm.) Britton & Rose var. *pubescens* Borg; also Y. Ito, *nom. nud.* No specimen or locality mentioned. *Mammillaria strobiliformis* Scheer var. *caespititia* Quehl. *Escobaria tuberculosa* (Engelm.) Britton & Rose var. *caespititia* Borg. "Heimat unbekannt, vermutlich Mexiko." *Mammillaria strobiliformis* Scheer var. *gracilispina* Quehl. *Escobaria tuberculosa* (Engelm.) Britton & Rose var. *gracilispina* Borg. No locality or specimen. *Escobaria arizonica* Hester, *nom. nud.* Type locality: north of Portal, Chiricahua Mountains, Cochise County, Arizona.

4B. Var. *Orcuttii*
(Rose) L. Benson, **comb. nov.** (p. 26)

The most central two or four spines shorter than the surrounding 4 to 6 central spines, several up to ⅜ or sometimes ½ inch long, 1/72 inch in diameter.

DISTRIBUTION: Limestone hills in grassland at about 4,000 feet elevation. Desert Grassland. Arizona in the eastern parts of Cochise County. In New Mexico from Hidalgo County to Doña Ana County.

5. *Coryphantha missouriensis*
(Sweet) Britton & Rose

Stems solitary or branching and forming clumps 2 or 4 inches high and 6 to 12 inches in diameter, dark green, hemispheroid to depressed-globose, 1 to 2 inches long, 1½ to 2 or 4 inches in diameter; areoles white-woolly when young; spines partly obscuring the stem, yellowish at first but becoming dark gray, *pubescent;* central spine(s) commonly none but sometimes 1 or rarely 2 per areole, not conspicuously differentiated; radial spines 11 to 20 per areole, spreading, straight, ⅜ to ¾ inches long, basally up to 1/72 inch in diameter; petaloid perianth parts yellow or sometimes pink or red or tinged with red, linear-lanceolate, ½ to 1½ inches long, 1/16 to 3/16 inch broad, sharply acute, acuminate, or attenuate, entire; fruit red, ⅜ to 1 inch long, of nearly the same diameter; seeds black, obviously punctate, helmet-like, 1/24 to 1/12 inch long, 1/36 to 1/12 inch broad; hilum clearly basal, slightly oblique.

5A. Var. **Marstonii**
(Clover) L. Benson, **comb. nov.** (p. 26)

Central spines none; radial spines 10 to 13 or 19; flower 1½ to 2 inches in diameter; sepaloid perianth parts fimbriate; seed 1/12 inch long.

DISTRIBUTION: Hillsides in the lower forest belt and in woodland at 5,000 to 7,000 feet elevation. Rocky Mountain Montane Forest and Juniper Pinyon Woodland. Garfield County northwest of Boulder, Utah, to the Buckskin Mountains (Kaibab Plateau) and the vicinity of Prescott, Yavapai County, Arizona.

SYNONYMY: Coryphantha Marstonii Clover. Type locality: Above Boulder in Garfield County, Utah. *Neobesseya arizonica* Hester. Type locality: Buckskin Mountains, Arizona.

Culture and Care of Arizona Native Cacti

Data for this section contributed by A. A. NICHOL

GENERAL CONSIDERATIONS

If a few simple rules are followed, Arizona cacti may be planted and grown successfully in Arizona, indoors or out.

Transplanting

In transplanting cacti, freshly cut roots should not be placed in contact with the ground. This is also true for cuttings. After a plant has been dug up or the cutting made, it should be put in an airy but shady place for several days until the cut surfaces have thoroughly dried or callused. This will be four or five days in summer and two or three weeks in winter. When the plant is taken from the ground, the roots should be cut back, leaving only enough of stubs of three or four of the heaviest roots to hold the plant firmly upright when it is returned to the ground. This rule is advantageous to follow even if the cuttings or plants are started in propagation benches where nothing but clean, washed sand is used. For starting young stock or handling small plants, a sand bed is helpful; if a sand bed is used, watering every other day is sufficient in the summer, and twice a week is adequate in cool weather.

Soil

For indoor cultivation in pots one of the most successful soils is obtained by the use of granite rock and sand. This neutral medium includes none of the injurious elements found in many of the other types of Arizona soils and rock. A good combination is a handful of coarse granite gravel in the bottom of the pot followed by 2 or 3 inches of good humus soil similar to that used for potting geraniums, *Coleus* or other house plants. The remainder of the pot should be filled with clean, coarse sand that has been mixed with 2 or 3 tablespoons of hydrated lime. Many other potting mixtures are explained in numerous books and special journals, such as the *Cactus and Succulent Journal.*

Cultivation

Cacti have shallow root systems. The roots may run in all directions for many yards, but many will be no more than 2 or 3 inches below the ground surface. As it is almost impossible to disturb the soil without injuring or cutting some of the roots, the only cultivation recommended is removing weeds and raking the soil surface for the sake of improved appearance.

Water

Although cacti may tolerate frequent watering, and their growth is partly proportional to available water, if free water stands about the root crowns for any length of time, many of the plants will be lost by rot. Therefore cacti, either in pots or gardens, should be elevated a few inches above the surrounding soil so water will drain quickly away from the stem bases.

In the fall, certain species native to the higher altitudes where winter temperatures are low go through a water-losing process which reduces the hazard of frost injury. This process has all the appearance of wilting, but the plant fills out and becomes entirely healthy upon the return of warm weather the following spring. The species having this characteristic should be known to the gardener, since the repeated application of water to the "wilted" plant at this time may be harmful. Some Arizona species exhibiting this dehydrating phenomenon upon the approach of cold weather are *Opuntia Whipplei, O. polycantha, O. macrorhiza, O. erinacea* and var. *hystricina, O. chlorotica, O. fragilis, Echinocereus Fendleri, E. fasciculatus* var. *Bonkerae, E. triglochidiatus* vars. *melanacanthus* and *neomexicanus, Neolloydia intertexta, Sclerocactus Whipplei, Coryphantha vivipara* vars. *bisbeeana* and *arizonica, Mammillaria Grahamii* and *M. Wrightii.*

Unless the gardener knows definitely that his plants are suffering from lack of water, a wilted specimen should not be watered. The chances are that something else is to be blamed, and the plant should be dug up and thoroughly examined. Frequently large boring insects get into the crown of the plant, or rot may enter at some point near the ground level.

Seedlings

All Arizona species can be grown from seeds, although it is often difficult to get the *Opuntia fulgida* group of chollas to sprout. If seeds are used, they should be as fresh as possible, because the older they become the slower they are in germinating. A satisfactory seedbed can be made in a shallow flat or other wooden box. One part of finely sifted garden soil should be mixed with three parts of screened sand. Ordinary 20-mesh window screen is satisfactory for screening. The soil should be settled evenly and firmly in the flat with a block of wood and the seeds scattered on the surface. Then additional soil should be sifted above the seeds to a depth of two or three times the greatest seed diameter. A piece of burlap should be fitted smoothly inside the flat, and water should be applied through the burlap. When the seeds start to sprout the cover may be removed. If the seedbed is made indoors and much moisture is present, a good insurance against losing the seedlings is the use of Semesan, a small packet of which can be purchased at any seed store.

SPECIAL GROUPS

Prickly Pear and Chollas

All the *Opuntias* grow readily from joints or cuttings, although in the tuberous-rooted species growth is much more rapid if the root is included.

While the soil range of prickly pears and chollas is broad and hard to delimit, a few species are restricted definitely to light, sandy soils. Among these are *Opuntia Stanlyi* vars. *Kunzei* and *Parishii* and *O. ramosissima.* If any of these three is desired in a garden or collection, it should be planted in a bed of nearly pure sand raised 8 to 12 inches above the surrounding ground. If the bed is gradually rounded to "feather" into the remainder of the garden, the plants will cover the mound in a dense mat and will be very effective. *Opuntia Stanlyi* also grows best in light soil, but it has a greater range of tolerance than the three chollas mentioned above.

Cereus

The genus *Cereus* ranges through a wide variety of soils, and the limitations are mechanical rather than chemical. Saguaros, for example, will do well in heavy to sandy soils provided they have anchorage. Their great weight and the leverage exerted upon the bases make them likely to be tipped over in storms if rains previously have softened the ground, unless the large "anchor" roots have an opportunity to become wedged among rocks. Unsightly guy wires and props in gardens can be dispensed with if two or three pieces of pipe are driven into the hole the saguaro is to occupy. Two or three of the stubbed anchor roots may be wired to the pipes, and the roots should be wrapped with a few turns of burlap where the wire goes around them. Plain black 9 or 12 gauge wire is desirable, and it will rust away by the time the plant is able to support itself. The same procedure can be used for other large cacti, such as the organ pipe and the senita *(Cereus Schottii)*. In transplanting saguaros, care should be taken not to injure the radishlike taproot directly beneath the column of the plant. If this is injured, the saguaro may stand for years without growth, or it may die.

The night-blooming cereus is best transplanted when the large turniplike tuber is planted. This should be well dried and callused before planting, and the slender stems should be given support.

Echinocereus

The genus *Echinocereus* is divided sharply between species that thrive in ordinary sandy soils and species that need an abundance of humus. *Echinocereus Fendleri* and *E. Engelmannii* are typical of the group more resistant to poor soils; *Echinocereus Fendleri, E. fasciculatus* var. *Bonkerae, E. triglochidiatus* vars. *melanacanthus* and *neomexicanus,* and *E. pectinatus* var. *rigidissimus* grow best in well-drained leaf mold above coarse gravel, and they require about one-third shade.

Mammillaria

A great diversity of requirements is found in *Mammillaria.* The common *Mammillaria microcarpa* can make the greatest adjustment, only neutral well-drained soil being required. *M. gummifera* var. *MacDougalii* needs part

shade and a porous or gravelly soil with a fair amount of humus. Care should be taken not to break the taproot. *M. Mainae* needs half shade, and it should be planted on a mixture of broken twigs and leaf mold. *M. Wilcoxii* and *M. Wrightii* also need part shade. *M. Thornberi* does best in a bed of pure sand 4 or 5 inches thick underlain with good garden soil. *M. tetrancistra* needs full sun, well-drained sand of granitic origin, and very little water.

Echinocactus

The Turk's head *(E. horizonthalonius* var. *Nicholii)* will do best in a shale soil or one that has been made up with about one-half lime.

Coryphantha

With one exception, the species of *Coryphantha* will grow under a wide range of conditions. The exception is *C. recurvata,* which will do best when planted in about half shade on a mound composed of about two parts of coarse gravel to one of leaf mould. *C. Scheeri* var. *robustispina* grows excellently in river silt with a small percentage of shade. Although *Coryphantha vivipara* var. *bisbeeana* is tolerant to sun and any soil, quality and vigor are much improved if the plant is raised humus bed of the type recommended for *C.* slightly above the ground level on a gravel-*recurvata.*

Index

Accepted names of genera and species in this index are in roman type, thus, 126; names not accepted as standard by the author are in *italic* type, thus, *126*.

Illustrations are indicated by page numbers in parentheses, thus, (126).

Major discussions or critical explanations of a subject are indicated by page numbers in roman type. Minor discussions are indicated by page numbers in *italic* type. No attempt has been made to list every page containing casual mention of a given subject.

Personal names are given only as they appear in text discussion or in reference to publications — not in association with botanical terminology.